LEADING
WITH
VALUE

LEADING WITH VALUE

How to Effectively Communicate Business Value

Peter B. Nichol

For information about permissions to reproduce selections from the book, or about special discounts for bulk purchases, please contact:

booksales@datasciencecio.com

First Edition

Publisher's Catalog-in-Publication Data
Library of Congress Catalog Card LCCN No. 2020922775

ISBN: 978-1-7363859-0-6
Ebook: 978-1-7363859-1-3

Printed in the United States of America.

To every leader who struggled to articulate business value.
Your struggle is over. Dance with me.

CONTACTING PETER B. NICHOL

Sharing my knowledge gives me pride. I treasure the time I spend working with aspiring leaders, executives, and teams to elevate their performance and introduce the principles of value-based management.

The motivation for the development of this book sprang from a realization that leaders are too focused on the structures of management, e.g., project-management offices and lean-agile centers of excellence. A critical element was missing—a hyper focus on business value. Leaders uniformly agree that we should focus on business value over the number of projects completed. The challenge is how to communicate value.

As a 4x author, contributor to CIO online magazine, and Six Sigma Master Black Belt, I wanted to share my approaches for capturing value in an easy-to-understand format with straightforward examples.

Leading with Value addresses the challenge of how to capture and communicate business value for individuals, teams, and organizations to demonstrate quantifiable impacts. It explains how to move from abstract concepts to concrete examples by showcasing quantified business value that executives will understand and appreciate.

I welcome your stories, insights, and any useful tools and techniques you have to improve how to create unique ideas and deliver with excellence. I guarantee I'll read all the letters and email I receive. However, due to the pure volume, I'm not able to respond to every request for information.

For leaders, teams, and organizations interested in engaging me for board advisory, consulting, or speaking, please contact me at OROCA Innovations:

Peter B. Nichol
Chief Technology Officer, OROCA Innovations
145 Barkledge Drive

Newington, CT 06111
E-mail: peter.nichol@datasciencecio.com
Twitter: @peterbnichol

For additional information on the book, *Leading With Value*, please visit the website:
www.datasciencecio.com/leadingwithvalue

For additional information on the book, *Think Lead Disrupt,* please visit the website:
www.datasciencecio.com/thinkleaddisrupt

For additional information on the book, *The Power of Blockchain for Healthcare*, please visit the website:
www.datasciencecio.com/blockchainbook

For additional information on the book, *Learning Intelligence*, please visit the website:
www.datasciencecio.com/li

ABOUT THE AUTHOR

PETER B. NICHOL (@peterbnichol) is a technology executive with 19 years of experience who's dedicated to driving innovation, digital transformation, leadership, and data in business. He helps organizations connect strategy to execution to maximize company performance.

Peter has led businesses through complex changes including the adoption of data-first approaches for portfolio management, Lean Six Sigma for operational excellence, departmental transformations, process improvements, maximizing team performance, designing new IT operating models, digitizing platforms, leading large-scale mission-critical technology deployments, product management, and agile methodologies, and building high-performance teams.

Peter has been recognized for Digital Innovation by CIO 100, MIT Sloan, Computerworld, BRM Institute, and the Project Management Institute. As Managing Director at OROCA Innovations, Peter leads the CXO advisory-services practice that drives digital strategies.

Peter was honored as an MIT Sloan CIO Leadership Award Finalist in 2015 and is a regular contributor to CIO.com on innovation. As Head of Information Technology, Peter was responsible for Connecticut's Health Insurance Exchange's (HIX) industry-leading digital platform, which has transformed consumerism and retail-oriented services for the health-insurance industry. Peter championed the Connecticut marketplace digital implementation with a transformational, cloud-based, SaaS platform and mobile application. He was recognized as a 2014 PMI Project of the Year Award finalist, CIO 100, and received awards for best digital services, API, and platform. He also received a lifetime achievement award for leadership and digital transformation and was honored as a 2016 Computerworld Premier 100 IT Leader.

Peter has a Bachelor of Science in Computer Information Systems from Bentley University and a Master in Business Administration from Quinnipiac University, where he graduated Summa Cum Laude. He earned his PMP® in 2001 and is a certified Six Sigma Master Black Belt, Certified SAFe® Lean Portfolio Manager, Certified SAFe® Agilist, Certified SAFe® Practitioner (SP), Certified ScrumMaster®, Business Relationship Management Professional (BRMP®), Certified Business Relationship Manager (CBRM®), and the first globally credentialled Master of Business Relationship Management (MBRM®) by the BRM Institute. Together we can change the world!

For more details, please visit
https://datasciencecio.com/about/

ACCLAIM FOR PETER NICHOL'S BUSINESS BOOKS

LEADING WITH VALUE

"We all know that traditional PMOs are broken and don't provide the value executives expect. *Leading with Value* is the first book to systematically break down the concept of value into pieces that every leader can take advantage of for their organizations' benefit. If you want to do a better job of quantifying the impact and value of your team, this book should be your first read!"

— TRUSHAR SHAH

HEAD OF ENGINEERING, GEP WORLDWIDE

"Learn how to speak the language of business value! *Leading with Value* is the solution that every business relationship partner has been looking for—the complete guide to effectively communicating business value!"

— GLENN REMORERAS

CHIEF INFORMATION OFFICER, MARK ANTHONY GROUP

"CIOs expect their leaders to understand how to quantify value. Nichol takes the mysterious and complex topic of value and simplifies it to the essential nuggets we all need to understand. Struggling to quantify the value of your team? Pick up a copy of *Leading with Value* and stand out from your peers with your understanding of how to effectively communicate value!"

— HATEM ELMANAWY

DIRECTOR OF PROJECT MANAGEMENT, OCEAN CAPITAL

"Value is, perhaps, the most misused word in the corporate lexicon. Identifying and aligning on what value must be captured is itself a challenge, and companies rarely follow-through during and post capability acquisition to ensure anticipated value is, indeed, captured as planned with little to no erosion. Quite admirably, Peter has taken on and delivered on this challenge to explain a complex topic in straightforward terms. *Leading with Value* provides a holistic take on value - Identification, Stewardship, and Realization - that brings us back to the basics while challenging our thinking."

— RAM MADRAS

EXECUTIVE DIRECTOR OF IT STRATEGY & PLANNING, BLUE CROSS AND BLUE SHIELD

"Nichol finally wrote the business book for all levels from the CIO to the manager. Every leader is charged with creating and capturing business value. *Leading with Value* takes the time to elaborate on how specifically to take abstract concepts and turn them into concrete and quantifiable examples of value. Change how you look at business problems forever with *Leading with Value.*"

— ISAAK GEORGIADIS

HEAD OF VALUE COST MANAGEMENT, BEI DRÄXLMAIER GROUP

"*Leading with Value* offers insightful examples of how to capture and communicate value to your business partners. Do your leaders understand your value? Can your peers describe the services you provide? As a CIO and 4x author, Nichol harnesses his experience to empower leaders to maximize business value!"

— KRIS SPRAGUE

HEAD OF PROJECT PLANNING & SCHEDULING, REGENERON PHARMACEUTICALS, INC.

"*Leading with Value* provides several methods and the essential nuggets that allow companies to transform Project-Management Offices into Value-Management Offices. Nichol's new book is a must-read for Senior Executives, PMO leaders and project managers that want to expand their knowledge on discovering, realizing and optimizing the value of their project, teams and portfolios!"

— JEAN BINDER

HEAD OF LIFE SCIENCES PROGRAM MANAGEMENT OFFICE, PHILIP MORRIS INTERNATIONAL

"*Leading with Value* presents a unique formula that helps leaders maximize business value. If you're a leader looking for methods, approaches, and solutions to quantify business value, this book will be your best guide!"

— NANJI CHANDRASEKHAR

VP OF IT INFRASTRUCTURE & CLOUD OPERATIONS, ASPEN DENTAL MANAGEMENT, INC.

"*Leading with Value* offers CDOs and CIOs concrete ways to capture and communicate the business value of their data, analytics and technology investments. Nichol makes the case for why traditional budget and PMO measures are outdated and offers more relevant approaches for a rapidly changing digital world. A must read!"

— CINDI HOWSON

CHIEF DATA STRATEGY OFFICE THOUGHTSPOT & HOST OF THE DATA CHIEF PODCAST

THINK LEAD DISRUPT

"*Think Lead Disrupt* presents unique insights on how innovators can continually evolve ideas for digital transformation. Nichol, a respected writer for CIO, has shared the secrets to move from strategy to execution. A must read!"

— AMRESH MATHUR

DIGITAL TRANSFORMATION EXECUTIVE, SAMSUNG ELECTRONICS AMERICA

"*Think Lead Disrupt* will illuminate how brilliant leaders identify new ideas to transform their businesses. Nichol presents the practical impacts of how ideas lead to disruptive business transformations. A refreshing look at how evolutionary and revolutionary ideas fuel business transformation. A must-read for every innovator!"

— QAMAR ZIA

SENIOR DIRECTOR OF TECHNICAL SERVICES, KÖRBER SUPPLY CHAIN

"A prolific writer in a Zen master fashion strikes again. Building on his CIO experience, Peter dissects the anatomy of innovative ideas and creativity once again in his upcoming book. *Think Lead Disrupt* is riveting, addictive, and disruptive. How leaders think, innovate, and grow from the ashes? If you are looking for the food for thought, the meal has been served."

— MINHAAJ REHMAN

CEO AND FOUNDER AT PSYDA

"*Think Lead Disrupt* is a fantastic book for leaders that want to develop innovative ideas that can be translated into world-class business products, services, and interactions. Nichol's voice as a CIO and global thought leader is simple to understand and easy to apply. This is a must read for innovators leading teams, departments, or organizations that must be on the cutting edge!"

— AARON S. BARNES

CEO AND CO-FOUNDER, BUSINESS RELATIONSHIP MANAGEMENT (BRM) INSTITUTE

"A wise man—my 100-year old grandfather—once told me, when faced with a complex problem, always go back to first principles. From there, assumptions can be questioned, and the status quo can be challenged. *Think Lead Disrupt* embodies this philosophy and presents unique insights into how innovators can continually evolve digital transformation ideas. Nichol, a respected writer for CIO, has shared the secrets to move from strategy to execution. A must-read!"

— ERIC ACTON

HEAD OF INNOVATION ECOSYSTEMS, ROLLS-ROYCE

"A must read. Everyone wants to think like an innovator. Every day we're immersed in a world that is lightning fast. In this world only the best ideas are funding, implemented and adopted. Nichol has captured why innovative executives are able to continually discover earth-shattering ideas."

— MATHIAS GOYEN, MD

CHIEF MEDICAL OFFICER EUROPE, MIDDLE EAST AND AFRICA AT GE HEALTHCARE

"The innovation mindset is fundamental to source new solutions. Without this you will not have the right foundations to really achieve true innovation. Peter clearly articulates how having the right innovation mindset provides a path to harness disruptive strategies and approach new ways to solve old problems!"

— STEVE SUAREZ

GLOBAL HEAD OF INNOVATION, FINANCE AND RISK AT HSBC

"A great book for leaders looking to create an organizational culture and the capacity to accelerate innovation. A must read book that explores the art and science of innovation!"

— KRISHNA CHERIATH

VP, HEAD OF DIGITAL, DATA AND ANALYTICS AT ZOETIS INC.

THE POWER OF BLOCKCHAIN FOR HEALTHCARE

"*The Power of Blockchain for Healthcare* presents a crisp and fresh perspective of blockchain tailored to healthcare. Nichol focuses on how blockchain will enhance the patient and provider experience and the technology we use to interaction with patients. A must read!"

— RAJA RAMACHANDRAN

FOUNDER, RIPE.IO AND FORMER MANAGING DIRECTOR, R3CEV

"*The Power of Blockchain for Healthcare* is an essential read for any seasoned leader, and with Peter's unique perspectives it's no wonder he is one of the most influential innovation voices on healthcare innovation. A must read."

— GUILHERME STOCCO

HEAD OF STRATEGY AND INNOVATION AT BANCO ORIGINAL

"Peter offers a new voice presenting fresh perspectives on how blockchain will transform the healthcare experience. *The Power of Blockchain for Healthcare* presents the unconventional potential utilization of blockchain beyond the financial sector and within an industry that is on the verge of a disruption."

— EMILIA POPOSKA

CO-FOUNDER AND CEO, BRONTECH

"Blockchain technology will fundamentally transform, disrupt and change our world much like the internet has done. All industry segments will be affected, and it will allow for fundamental shifts in business models. Nichol magnificently articulates the implications for the healthcare industry. A must read for every senior executive who wants to be prepared for the future."

— JACK SMIES

VICE PRESIDENT & BUSINESS HEAD MIDDLE EAST AT WIPRO

"Peter accelerates past the typical explanation of blockchain, and immediately carves out blockchain's value for healthcare. If you're already familiar with blockchain but truly want to understand how blockchain will disrupt healthcare, then *The Power of Blockchain for Healthcare* is the book for you."

— ERAN ORR

FOUNDER & CEO AT VRPHYSIO INC.

"*The Power of Blockchain for Healthcare* provides a compelling argument for the transformative impact that blockchain will have on the patience experience. Nichol magnificently articulates a future vision of health, a vision every executive should experience."

— SAN BANERJEE

HEAD OF CONSUMER DIGITAL SOLUTIONS, HUMANA

LEARNING INTELLIGENCE

"Peter dives into unexplored, valuable territory in *Learning Intelligence*. A must-read for any leader wanting to compete in the innovation-powered landscape of today."

— MARSHALL GOLDSMITH

THE AUTHOR OF THE #1 NEW YORK TIMES BESTSELLER, TRIGGERS

"An important book for the new digital economy which requires fundamental skills that can adapt to a dynamic and fast moving environment. Peter has captured the quintessential element for preparing organizations to meet and exploit these digital capabilities."

— DALE DANILEWITZ

CHIEF INFORMATION OFFICER, AMERISOURCEBERGEN

"This is a great book, offering much wise advice. I most appreciated Nichol's insight around the value of data, information, knowledge and wisdom. *Learning Intelligence* is the new currency for exponential thinking."

— DAVID M. JOHNSON

SVP & GLOBAL HEAD OF PUBLIC SECTOR GROUP, COGNIZANT

"The ability to unlearn and learn is vital, as business leaders constantly navigate through change. *Learning Intelligence* is the tool they have been waiting for!"

— GURMEET CHAHAL
SVP AND HEAD HEALTHCARE, HCL TECHNOLOGIES

"Nichol is an insightful innovator and partner in driving transformation change within organizations. This book is a must read for business leaders charged with transforming their organization's ability to learn."

— LENNOX BROWN
CHIEF INFORMATION OFFICER, ANALOGIC

"Using well-presented research, Nichol makes a convincing argument for corporate leaders to grow their knowledge in this area. This book is a useful tool in recognizing the strategic impact of *Learning Intelligence.*"

— KARL PANSEWICZ
CEO, SNAPCODR, INC.

"Peter has identified the "secret sauce" for organizations to evolve their culture of innovation and learning. *Learning intelligence* is the missing link most organizations seek but rarely find."

— CHRISTOPHER CORNUE
CHIEF STRATEGY & INNOVATION OFFICER, NAVICENT HEALTH

"*Learning Intelligence* identifies a phenomenon that has defined this and future generations potential to advance and succeed – the ability to learn. We observe more and more that multidisciplinary approaches are key to finding new innovative solutions to challenging problems – and this requires all stakeholders to possess the learning agility necessary to make breakthroughs. I applaud Peter Nichol for capturing this in an easy-to-understand book."

— GEORGE T MATHEW, MD

CHIEF MEDICAL OFFICER, HEWLETT PACKARD ENTERPRISE

"Peter B. Nichol does an excellent job taking the important and complex, learning intelligence concepts and giving every individual advice to put into action to improve their ability to learn. *Learning Intelligence* is a great combination of strategy and practical approaches, sure to help anyone improve their ability to learn and innovate successfully."

— NICOLAS BRAGARD

CHIEF INFORMATION OFFICER, ESKER

"Peter is an expert at taking organizations from Information Strategy to Implementation to Value Realization. In L*earning Intelligence*, he applies this expertise to help individuals and organizations raise their Human Capital to new heights!"

— RICHARD SINGERMAN, PHD

ADJUNCT ASSISTANT PROFESSOR, JOHNS HOPKINS SCHOOL OF MEDICINE

"Peter's new book *Learning Intelligence* captures the essence of the "ability to learn" – the foundation for success. A must read, for all leaders who want to be successful in life!"

— BALU MV

CHIEF INFORMATION & INNOVATION OFFICER, AVA INTERNATIONAL

"It has been said that 'minding your P's and Q's is applicable to life and business. Peter Nichol's extraordinary new book enhances complete understanding of the Q's which is half the battle in successful startups and corporate enterprises. This is a game-changer."

— EDWARD BUKSTEL

CEO, CLINICAL BLOCKCHAIN

"A highly accessible book for all business executives that demand the best talent to meet operational objectives. *Learning Intelligence* provides a broad understanding of the ability to learn in a model that works brilliantly for both industry, government, and academic settings."

— SAFDAR ZAMAN

HEAD OF IT STRATEGY AND GOVERNANCE, NAKHEEL PJSC

"An important read for all leaders as we move towards a future, where speed, adaptability, knowledge, and sound decision making are the currency for success."

— STEVE LAWLER

SENIOR VICE PRESIDENT REGIONAL GROUP, CAROLINAS HEALTHCARE SYSTEM

"In *Learning Intelligence*, Peter B. Nichol launches a fascinating new concept with the potential to help business leaders significantly advance product and process innovation in every industry... The book clearly describes how leaders can acquire these valuable skills by accurately improving their *Learning Intelligence* and then continually working to measure their LQ."

— ROBERT COLI, MD
FOUNDER & CHAIRMAN, DIAGNOSTIC INFORMATION SYSTEM COMPANY

"Peter extends the current view on personal maturity by introducing LQ – Learning Quotient – a complement of intelligence and emotional intelligence quotients by focusing on our ability to change, grow and live. A fascinating view and an area which every leader must explore."

— THOMAS WALENTA
BOARD OF DIRECTORS AND FELLOW, PROJECT MANAGEMENT INSTITUTE

"Taking a new approach to intelligence and how it is best applied in your business Peter Nichol's new book *Learning Intelligence* will expand your mind and challenge conventions. We live in a new digital age of immediate and ready access to information that requires people to learn faster to match the exponential nature of change in our world. *Learning Intelligence* helps you rethink your world, business, teams and employees!"

— NICK VAN TERHEYDEN, MD
CHIEF MEDICAL OFFICER AT NTT DATA, INC.

TABLE OF CONTENTS

PREFACE

How do you communicate the value of yourself? Business executives won't understand your importance unless you can clearly explain your value. This also applies to the teams you lead and the organizations you support.

We, as leaders, struggle with quantifying value. Portfolio directors are searching for methods to reframe the benefits their teams produce. Program and project managers look for ways to communicate the usefulness of completed work. Business analysts model and remodel processes and explore methods to explain their impact. Every role from DevOps to portfolio executive has a common goal—how to effectively communicate business value.

Business value is difficult to grasp, so allow me to ground the concept by providing the alternative view—the absence of value. You're excited about the upcoming birthday of a close friend. You have a few ideas in mind to commemorate the occasion. With the birthday less than a week away, your options are limited. You log into your Amazon account and find the item you want. You order the $20 item and, upon inspection of the detailed invoice, discover a $40 fee. Irate, you call customer service, and you're informed that the fee reflects the cost of the concierge. It sounds reasonable on the surface, but the kicker is, you never used a concierge during your checkout process.

The customer-service agent explains that you *could* have used this service, even though you didn't. Regardless, Amazon has to pay for the concierge overhead, and the charge stays. How do you feel?

This is the exact feeling every business partner has when they're forced to pay for resources they don't want and didn't ask for. This is how your project-management office (PMO) business partners likely feel today. Business value eliminates that pesky concierge fee and offers hyper-transparency of the value the service offers. In our example, the customer-service agent would have informed you before your purchase that if you chose concierge service, you'd have your product in two business days versus five to eight business days using self-service. Without that transparency, you can't quantify the benefit of that service offering and make the proactive decision to accept or reject use of the service.

Throughout my career—leading teams and organizations and supporting demanding executives—the most challenging question has remained constant: How do you quantify the value of your team?

Portfolio directors want to increase productivity of the team. Program and project managers push to deliver initiatives on time. Business analysts dive into the business details searching for alternatives. The problem is, none of them know where to find the value.

Leading with Value is the guide they never had and always wanted. Through my experience building world-class, high-performing teams, I've discovered how to capture and communicate value.

In this book, I explore the factors of how to effectively communicate business value. The book has five chapters:

Chapter 1, Value Management as an Organizational Capability, uncovers the root causes of why traditional PMOs are broken. We jump into the concept of a value-management office and discuss the powerful mindset shifts that occur when leaders unplug ineffective PMOs. We adventure into four of the world's most impactful value-management approaches. We wrap up our discussion by identifying dozens of alternative value-management approaches that can be leveraged to quantify value.

Chapter 2, Applying Techniques to Quantify Outcomes, walks us through specific business cases and how they connect abstract concepts to quantified value. We apply techniques to capture value across four common delivery situations: project delays, assessing project managers, evaluating vendor performance, and reining in financial overspending. Chapter 2 concludes with a discussion of useful techniques to capture value depending on the context.

Chapter 3, A Step-by-Step Guide to Articulating Value in Project Management, is a practical guide to quantifying value. This exercise takes the reader through a detailed explanation of what, specifically, is required to quantify impact and articulate value. Through this example, the reader will fully understand what is meant when we say, "Articulate the business value." After reviewing improvements metrics, process mapping, waste identification, value-stream mapping, and financial savings, we shift quickly into Chapter 4.

Chapter 4, Assessing Project Management Competencies, explores why defining "good" is essential for team growth. We identify which functional areas require focus for assessment and evaluation, and explain the powerful concept of demonstrated competencies. With the standard for good defined and with our assessment completed, we share with the reader an example of crisp feedback for individuals and teams. Before exiting this topic, we also cover organizational project-portfolio management maturity.

Chapter 5, Leveraging Service Catalogs to Communicate Value, brings everything together and shares why it's vital that your business partners understand what you and your team offer the organization. Business leaders aren't too thrilled when they incur charges that are unplanned and the value they provide is unclear.

Leading with Value will change the way you think, lead, and communicate with your leaders. Don't delay. Amplify your impact today!

Peter Benton Nichol
Connecticut, USA

ACKNOWLEDGMENTS

The author would like to acknowledge the contributions of numerous anonymous reviewers from the finance, healthcare, pharmaceutical, biotechnology, energy, and retail industries and those from the value-management community who reviewed and shared insights around the concepts of value-management.

"Any powerful idea is absolutely fascinating and absolutely useless, until we choose to use it."

– RICHARD BACH

CHAPTER 1

VALUE MANAGEMENT AS AN ORGANIZATIONAL CAPABILITY

"Price is what you pay. Value is what you get."

- WARREN BUFFETT

Chapter Objectives

After reading this chapter, readers will be able to:

- Explain the four most impactful approaches for value management
- Identify multiple metrics to observe, define, capture, realize, and optimize individual, team and organizational value
- Understand how to apply the Value Quantification Value Framework
- Describe use cases for the application of the Rapid Economic Justification value management concept
- Know which metrics are more effective based on the value management framework selected

The Shift to Value-Management

This chapter aims to present a practical approach to institutionalizing business value realization as an organizational capability. Business and technology executives are under continuous pressure to justify investments and validate business outcomes.

When executives are asked to explain the value generated by their team, department, or company, only then do they realize they lack the in-house expertise to produce the level of quantifiable outcomes expected. Progressive leaders know that the process of discovering, realizing, and optimizing value is an essential organizational capability. This chapter presents a practitioner approach offering multiple value management frameworks to assist executives in making informed decisions when building a value-management office.

Traditional project-management offices maintain project-management standards, establish best practices, define common languages, develop a resource-management view, and create and maintain project artifacts and tools. Unfortunately, the heavy burden of these core project-management office activities results in no time for strategic planning. The value-management office oversees the execution of all the company's strategic programs. The company's hyper-focus on connecting strategy to execution ensures the value maximization of its strategic initiatives.

A Value-Management Office Ensures Projects Achieve Value

Project-management offices measure success based on traditional key performance indicators, but these don't align with value. It's essential to ensure that project-management standards are in place and that best practices are being used to manage and monitor project success. There's just one problem—project completion doesn't equal value maximization. In other words, it's absolutely possible to deliver a project that has effectively no value. For example, a business application could have been developed and launched into production on the agreed schedule. Normally, this would be deemed a success by the project-management office. However, what if there were 100 users, and all 100 were still using the old system? Was the project delivered on time and on budget? Yes. If all 100 users are still using the old application, did the project provide any value? No. This is the difference between a project merely being completed and a project achieving the maximum value. On the other hand, if this project did, in fact, have 100 users shift from the old to the new system, that would indicate—assuming all other factors are consistent—that the maximum value was achieved.[1]

The benefit of constructing a value-management office ensures that strategic programs achieve their value. Progressive companies are spinning up value-management offices to oversee all the company's strategic programs to monitor and measure not only program completion but value realization.[2]

Once executive leaders buy in to the concept and benefits of a value-management office, the real challenges begin. Who's qualified to run this new office? What's the focus of the office? Which framework do we adopt, or do we build our own value methodology? Is the office confined to a vertical, or does this responsibility span the entire organization? These are all valid questions that need to be answered. One of the most challenging questions is to determine which approach the company will use to capture and report on value.

Typically, there are three main approaches to value management: (1) qualitative, (2) quantitative, or (3) mixed. The qualitative methods focus on soft benefits such as peer reviews, feedback, or case studies. The quantitative methods focus on hard benefits such as payback period, total economic impact, or net present value. The mixed methods combine both the qualitative and quantitative methods in entities such as a balanced scorecard, Silk's method, or the PENG model.[3]

Whether you choose a qualitative method or a quantitative method, you inevitably address the needs of some stakeholders and omit the needs of others. This results in the sponsor being extremely happy with the results, while other important stakeholders aren't satisfied. The answer, of course, is finding a balance. Choosing a net present value might be a great choice if reporting to a chief financial officer. Alternatively, if that team is reporting to the chief marketing officer or the chief science officer, those metrics won't hold much utility.

When selecting a value-management approach, the mix of sponsors and stakeholders will directly influence which value approach is more applicable for your business. It's vital to choose a value-management approach that's fit-for-purpose.

The research presented here assumes a solid knowledge of the fundamentals of project management, knowledge of executive reporting, understanding stakeholder analysis, evaluating critical success factors, and general awareness of the delineation between delivering a project and delivering value. These topics and many more have been covered more thoroughly in previously published research.

The Four Most Impactful Approaches to Value Management

Our research focuses on the four value-management approaches that are the most effective to roll out organizationally with the greatest near-term impact: (1) Rapid Economic Justification, (2) Value of One, (3) Value Quantification Framework, and (4) the Before-and-After Model. Each value approach will be elaborated on with an explanation of what the model is, why it matters, and how to apply it generally.

We'll begin by covering each of these value-based models in detail and explain how they can be applied organizationally to capture value.

The Rapid Economic Justification (REJ) model balances the cost and profit potential of a project to determine its value. This model considers critical success factors (CSF) and conducts business exercises to evaluate IT investments. This model is best applied to individual projects or investments. The general model has five steps. First, conduct an assessment of the business situation. Second, define the solution. Third, perform an estimate of the profit and loss (P&L). Fourth, define the risks. Fifth, calculate financial key performance indicators.

The Value of One (V of 1) model is a simple, quantitative technique. The model focuses on the value of a solution by isolating and quantifying a single observation. Often, when the value is presented as a net amount or in cumulative form, the aggregated numbers are difficult for people to grasp. This method narrows the focus to a single observation and channels the discussion to reach an agreement on the impact of a single event. This is one of the simplest and most powerful concepts in value management, because the output is extremely relatable to stakeholders who are both familiar and not familiar with the initiative.

The Value Quantification Framework® is a framework I developed as a foundation for starting a value-management office.[4] The concept centers around team collaboration and consensus to lock in organizational alignment. This model has five key steps: value observation, value definition, value capture, value realization, and value optimization. This model allows for a playbook-type introduction with a low barrier for organizational adoption. This framework isn't typically applied to all projects but rather to strategic projects that require maximization of value and the quantification of outcomes.

The Before-and-After Model is another pragmatic model that focuses on simplicity over intricate financial expression to present value to stakeholders. This model leverages a single holistic view, illustrating the process before and the process after. While elementary, this is a great method for vendors to demonstrate delta value between the current state and the future or realized state. The presentation of this model on a single slide or view is especially impactful. This model can be particularly effective when results or outcomes are shown to a diverse group of stakeholders.

For consistency of delivery, as we review each value model and apply it to a business context, we'll explain each model with the following four elements (an abridged scientific method) for simplicity:

1. Opportunity: the question we're asking
2. Description: the intent and any potential limitations of the value model
3. Benefits: the primary advantage of using and applying the model
4. Metrics: a brief list of common metrics that are complementary to the model

This format will ensure the models' application can be consumed by a wide range of readers, observers, and leaders.[5]

Rapid Economic Justification

The Rapid Economic Justification model is a thoughtful approach to optimizing IT investments to establish alliances between IT and business. This model frames the project in terms of business-performance improvement and links the initiative to organizational critical success factors.[6]

OPPORTUNITY

The business case is that a business is developing and deploying a new application to manage the functions around human resources including staffing, development, compensation, safety and health, and employee relations. Business partners are unclear on the impact that the final solution will have on people, processes, and technology.

DESCRIPTION

The Rapid Economic Justification model has five core steps. First is the business assessment (identify stakeholders, identify critical success factors, identify key performance indicators, identify key business strategies, identify activities, and assess risks). This step reviews and validates the business need and establishes the base for business approach and justification. Second is mapping solutions (prioritize activities and process, identify opportunities for improvements, identify technology enablers, and assess risks). Here the focus is on evaluating each alternative for strengths and weakness and their impact on results. The third is the cost-and-benefit analysis (analyze and maximize benefits, quantify benefits, identify costs, quantify costs, and assess risks). In this step, direct and indirect costs are calculated against projected revenue. The fourth is to estimate risks (identify risks, manage the risk register, and create risk statements). This step conducts a detailed analysis of risks and the mitigation plan for each. The fifth step is building financial metrics and developing a value proposition (prepare financial metrics, prepare an economic justification with strong value propositions, present findings, and secure funding for the initiative).[7]

BENEFITS

This model does an excellent job of linking strategy to KPIs of the current state to KPIs of the desired state to business activities and processes. This approach is incredibly impactful when a team requires a high degree of collaboration to secure buy-in for a solution.

METRICS

This model's flexibility enables a wide range of quantification in KPIs reported. Seven metrics might include:

1. % of the manual and paper-based workflow
2. % budget reduction year-to-year

3. Number of data-quality verifications and conversions
4. Number of new services created
5. Number of complaints divided by total new orders
6. Number of average days for a new order
7. Number of handoffs per event avoided

These metrics will vary depending on the industry and business domain. The model is being developed to be amplified. However, this process effectively links stakeholders to the new business processes in a single-table view:

1. Stakeholders
2. Critical success factors
3. Key performance indicators
4. Strategy
5. KPI current state
6. KPI desired state
7. Business activities and processes

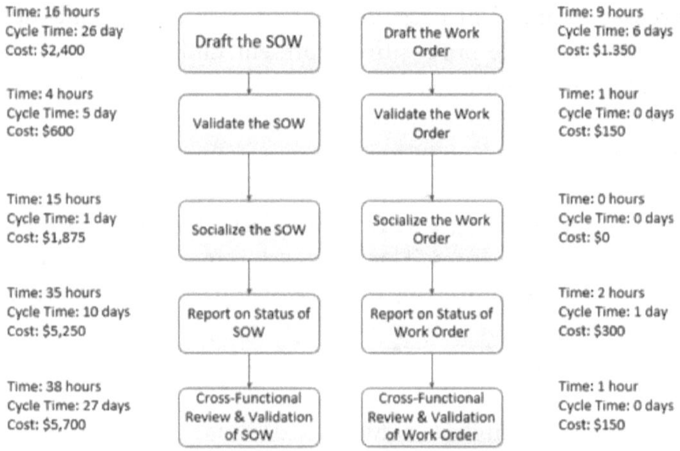

Value of One

The Value of One is an effective value method to quantify value. The technique is similar to other approaches such as expected monetary value, real options, goal-seeking, and leveraging benchmarks for specific scenarios. The objective of this method is to filter down the benefits into a single and individual observation.

OPPORTUNITY

The business case is that a team is processing contracts slowly, and there's a backlog, which is quickly building frustration among business partners. Not much is known about the process. However, it's evident that supply (contract-processing time) isn't meeting demand (the number of contract requests per week).

DESCRIPTION

The Value of One takes into account the time (activity effort), cycle time (duration), and the cost (internal and external expense) for a single observation.

This method is extremely useful for unifying leadership that doesn't agree on anything. By not speaking in generalities and selecting a specific observation, we don't refer to history or a theoretical culmination of benefits. This makes this technique impactful and real for stakeholders.

The benefit value is calculated using the formula below:

Benefit value = [value of 1 KPI unit] x [target number of KPI units]

BENEFITS

The model does a solid job of breaking a complex process down and managing those pieces. Once agreement is achieved on the individual observation, rolling up the value and benefits is a straightforward exercise.

For example, below is a simple Value of One model of a single observation. The left side represents the current process, and the right represents the reengineered process.

The illustration highlights the difference between the initial process (total time of 108 hours, 69 days cycle-time, and cost of $15,699 per contract) and the reengineered process (total time of 13 hours, seven days cycle-time, and cost of $1,950 per contract).

The result was ~$14,000 avoided per contract and an 89% reduction in cycle time. This type of quantified value is straightforward and easy to communicate.

METRICS

This model has similar traits to Lean Six Sigma approaches in terms of removing waste and to Six Sigma's focus on defect reduction.[8] Seven useful metrics might include:

1. Total effort
2. Cycle-time
3. Cost avoided
4. Touch time
5. Value-added time
6. Core processing time
7. Wait times (mean and deviation, or distribution)

Metrics focused on process reductions typically lend themselves well to this model.

Value Quantification Framework®

The Value Quantification Framework® is a method for evolving a team from an abstract concept of value to defined measures of how value will be measured and controlled. This model leverages the best of many models into a combined framework that's flexible and ensures the essential problem is solved first.

OPPORTUNITY

The business case is business leaders don't have adequate data to make informed and intelligent decisions. As a result, the data's confidence is low, given the poor data quality and limited business rules for ingesting data. The team is looking for a solution to tighten the data controls with measurable metrics that will ultimately demonstrate quantified value realized.

DESCRIPTION

The Value Quantification Framework® is a team-based framework to adopt value-management techniques within an organization. This model has five core steps: value observation, value definition, value capture, value realization, and value optimization. Value observation identifies the statement of need or problem statement from a business perspective. Value definition identifies and documents the current and future state of the situation. Value capture collects tangible and intangible (hard and soft benefits) elements of value. Value realization confirms that assumptions made are real and conducts an evaluation of the impact the initiative had on value. Lastly, value optimization revisits our known variables (current and future state, risk, and KPIs) to validate their improvement.[9]

BENEFITS

The model serves as a great framework for new teams that don't have experience articulating value or presenting quantified impacts to their business partners. Making the process mechanistic helps to ensure that all the critical steps are completed throughout the process. This is especially useful when there are multiple teams involved throughout the process (architecture, quality, development, business development, legal, sourcing, etc.).[10]

METRICS

As with the other models, the metrics will vary greatly based on what type of initiative is being modeled. However, seven relevant metrics might include:

1. User adoption
2. Data-entry errors
3. Number of active users
4. Business-satisfaction surveys
5. One-time cost savings
6. Bottlenecks removed from the process
7. Number of groups supported

When considering metrics, it's important to consider your audience. If you're speaking to a financial controller, KPIs such as costs avoided or dollars of capital spent might be more useful. Alternatively, if you're speaking with the chief operating officer, they might be more interested in headcount reductions or product-throughput improvements. Tailor the metrics to your audience.

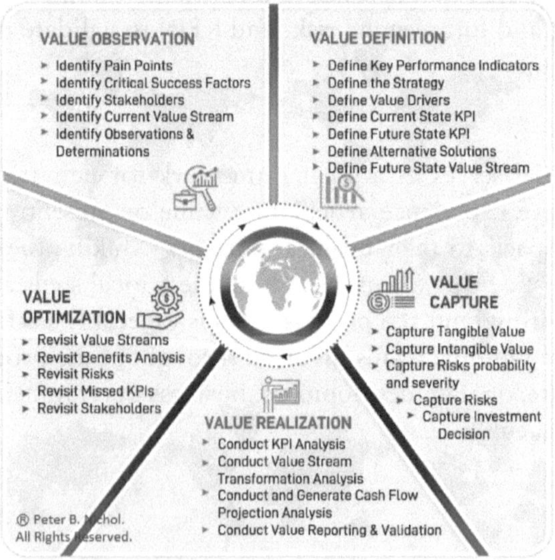

Before-and-After Model

The Before-and-After Model is more of a concept than a model. The concept is to show a dramatic difference in the "before" versus the "after" state of an initiative. By visually displaying the before and after state, it's clear to the stakeholder how progress was advanced.

OPPORTUNITY

The business case is that we're implementing a project-management approach for project-manager optimization. This is the second or third attempt to introduce portfolio-management optimizations after existing leadership failed to deliver on expected objectives.

DESCRIPTION

The Before-and-After Model is best applied when the progress made is either vague or insignificant. The model begins by explaining the current state—what happens today. The more the current state (before) description can be quantified, the better. Next, the future state (after)—the realization of the initiative—is documented. The net result is a clear picture of the state of the union at the beginning of the initiative and where we stand today, making progress from the current to future state easy to understand.

BENEFITS

The model can be applied in the most diverse situations, but it works best when the delta or difference between the before and after might be hard to extract from a written narrative.

When comparing processes, the Value of One model can prove useful when applied in combination with this model. Regardless, the act of documenting the before and after makes identifying the delta more direct.

METRICS

The beauty of this model is that almost anything can be included as metrics. It makes the comparison more balanced and uniform, as the same metrics are applied for the before and the after column.

Seven metrics that can be compared using this method include:

1. Number of days processing
2. Dollars of cost savings
3. Number of handoffs
4. Number of teams involved
5. Number of interaction points reduced
6. % of process overhead
7. % productivity increase

Restructured the Portfolio Mgmt. Delivery for Optimization While Increasing Output

Streamlined and automated delivery processes to reduce the PM burden to unlock PM productivity benefits.

BEFORE	AFTER
• Lengthy on-boarding time • Complex delivery structures • Numerous project reporting data quality errors • Inefficiency resulted in excess PMs required for delivery	• $3.2MM more projects delivered due to optimizations • 306% PM productivity increase over Clinical Dev. • 74% more efficient than G&A • 26% more dollars managed, per PM

By applying these metrics, we start to collect useful data points to play back the value realized for a specific initiative.

While this is far from an exact science, it's useful in providing directional information as to progress made.

Areas to Explore to Accelerate Project Value Realization

The purpose of the following propositions isn't to theorize on conceptual possibilities based mostly on anticipated outcome predictions but rather to identify next-generation areas of future work that are worthy of exploratory efforts. I developed these propositions based on my research and potential applications to accelerate project-value realization and the positive impacts of organizational operational excellence.[11]

The qualification and quantification of value requires a flexible approach. Effective value management requires strong analytical skills. Commonly, the most effective leaders driving successful value-management offices have the following essential skills:

1. Six Sigma
2. Portfolio management
3. Relationship management
4. Statistics and mathematics
5. Learning intelligence

These five skills help to ensure that leaders are able to not only lead but take an active and hands-on role in guiding teams to articulate the value in terms that sponsors, and stakeholders understand.

Three propositions (P1, P2, and P3) are offered that have the potential to introduce impactful methods for absorption into value management.

PROPOSITION 1: ORGANIZATIONAL CHANGE MANAGEMENT TO FUEL AGILE DELIVERY

Many organizations are running programs and projects using agile methodologies under the premise that agile methodologies bring speed and agility to project delivery. This can be true. However, a missing and vital element is how organizational change management is overlaid on existing agile methodologies.

There are three major classifications of organizational change-management leaders: (1) industrial or reengineering leaders focus on the company as a machine (Juran, Taylor), (2) management and leadership theorists have a consulting mindset of coming into a business and fixing a problem (Mintzberg, Drucker, Kotter, Hamel), and (3) sociology and psychology change theorists focus on the human side of change (Cummings, Lewin).[12]

Prior models do cover important factors in change. For example, Lewin (unfreeze and freezing change), Womack and Jones (hyper-focus on lean customer experience), Kotter (urgency and the eight stages of change), Hamel (building a coalition for change), Luecke (driving momentum of change with a shared vision), Chen (EVA model and redesigning the business system for change), and Bevan (the seven factors stressing clarity of change and building a strong business case).

One of the most powerful models for organizational change management came out of Belgium and was published by KPMG in 2017. This model identified five steps that are essential for organizational change management:

1. Make it clear.
2. Make it known.
3. Make it real.
4. Make it happen.
5. Make it stick.

Applying this model, we can include key elements of change into each step:

1. **Make it clear:** set the urgency and define the team.

2. **Make it known:** set the vision, review the current state, review the future state, tag the gaps, get buy-in, and solidify communication.
3. **Make it real:** define training and training materials, identify culture shifts, identify impacted business processes, and validate role alignment.
4. **Make it happen:** find a quick win and implement a feedback loop.
5. **Make it stick:** create a transition plan, identify ownership shifts, define post-sprint communications, and post measures and metrics for sustainability,

By taking this organizational change-management model and overlaying it on top of a normal series of sprints, agile has a new foundation for executing complex organizational change initiatives successfully.[13]

PROPOSITION 2: VALUE-MANAGEMENT OFFICE TO COMPLEMENT THE PROJECT-MANAGEMENT OFFICE

Project management offices (PMOs) provide a critical function for companies by, ideally, creating transparency in how work is delivered. The type of PMO makes a difference in the guiding principles and, therefore, the scope of the PMO's responsibilities within an organization. Generally, there are three types of PMO: supportive, controlling, and directive. A supportive PMO is advisory in nature and primarily offers a consulting light touch. A controlling PMO (the most common) focuses on enterprise compliance to policies, procedures, standards, and guidelines for the company. The third type of PMO is directive. A directive PMO actively takes control of projects and manages these projects directly. These are all required and necessary functions for a healthy company to deliver projects.[14]

However, while the PMO is buried in the weeds of template compliance, cohort meetings to discuss process steps, and tactical project risks, the bigger picture is forgotten. The project delivery becomes a focus, but the value realized is not.[15]

PMO leaders aren't equipped or trained to develop a value-management organization. As a result, a new breed of strategic oversight is now being demanded by CEOs and CIOs to ensure that the most critical organizational and strategic initiatives are delivered with excellence and with the maximum amount of value possible.[16]

Innovative leaders will soon realize this logical separation between operations and maintaining a traditional PMO function and the strategic need for a value-management office that's hyper-focused on guiding programs to deliver the maximum value for the organization.

This change will be uncomfortable and require an executive who's willing to replace existing leaders with leaders that have different skills. Imagine a discussion with executive leaders in which the value realized over the previous six months is clear and aligned with business partners. This is the future of the value-management office.[17]

PROPOSITION 3: OPERATIONAL-EXCELLENCE REPORTING TO THE CIO AS A CORE CAPABILITY

Operational excellence is simply executing a business strategy more consistently and reliably than your competition. The proof of the success of operational excellence is in the results.

Operational excellence is often envisioned with lean manufacturing. If you had that initial thought, you'd be correct. Interestingly, operational excellence has moved quickly from the manufacturing floor into modern-day offices. It's usually not called operational excellence except by knowledgeable practitioners. You might have heard of DevOps, DevSecOps, process improvement, continuous improvements, strategic alignment, benefits alignment, process-waste removal, business-process management, or business-process reengineering. These are all forms of operational excellence.[18]

The difference between just improving a process and operational excellence is that, with operational excellence, the improvement, gain, win, or advancement is quantified. It's a measured improvement that's visible.

To be most effective, this function also requires heavy business analysis, process reengineering, and Six Sigma skills. Traditionally, these aren't found in your PMO.

New Age leaders will quickly identify the differentiation of this skill when compared to traditional PMOs and will staff a new leadership team accordingly. Getting to 10x results takes a calculated risk, and it also demands an understanding that, with scale, there's benefit in specialization.

Limitations and Direction of Future Work

First, this chapter contributes to the limited body of knowledge directly attributed to advancing the concept that value realization is a key strategic organizational capability. The application of value-management principles advances companies and can be limited by the internal knowledge of value-based frameworks. It's amazing the spectrum of value-management models that has been developed over the years, each addressing a specific need.

The models I selected to expand on have the greatest potential for growth and systemic adoption for portfolio executives. However, there are also many other models that address certain niches with regard to how to make informed decisions.

While I stand behind the four selected models that have been elaborated above, I also appreciate that readers may find utility in the various other models that exist. I acknowledge that there are limitations to the models I've presented.

With that context, below is a list of additional value-management models that can be applied to articulate the value of programs, projects, or organizational initiatives. The following value-management models could be explored in greater depth and applicability to determine how they could be applied to identify, capture, and report on value realized.

1. **Total Cost of Ownership** (TCO) is a method to capture all the costs of an asset over its lifetime including operating costs—e.g., an analytics subscription for data ingestion might be $15k one time but have reoccurring yearly costs of $55k. This model factors in total costs when making a purchasing or investing decision.[19]
2. **Total Economic Impact** (TEI) is a model that focuses on cost, benefit, and flexibility to evaluate risk—e.g., making a determination between two point-of-sale systems.[20]
3. **Rapid Economic Justification** (REJ) is a model presented by Microsoft to develop a business assessment roadmap (critical success factors and key performance indicators) to highlight how technology can influence success by using cost-benefit analysis, risk assessments, and performing standard financial analysis—e.g., rolling out a single complex organizational change initiative.
4. **Total Cost of Investment** (TCI) is a model to capture all the costs of investment including broker commissions, taxes, licensing, and any fees related to the transaction—e.g., for a company considering acquiring a smaller bioinformatics company for $200MM, this model would help reflect the cost of investment that far exceeds the initial price of $200M.
5. **Total Value of Opportunity** (TVO) is a model originated by Gartner in 2003 that's different from the total cost of ownership because it considers the benefits of alternative investments—e.g., if a company is considering signing agreements with cloud provider A or provider B, this model helps to explore alternative solutions that are viable financially.

6. **Balanced Score Card (BSC)** is a model that links initiatives to organizational strategic objectives ensuring energy is spent on enabling the strategic vision—e.g., if a portfolio director has a list of 151 initiatives that are all deemed important, this model can help determine which initiatives are truly advancing the company's strategic goals.
7. **IT Scorecard** is a model that links IT-specific initiatives to organizational strategic objectives to ensure energy is spent on enabling the strategic vision—e.g., if there's only funding to either expand the network infrastructure or deploy wireless access points throughout the building, this model can help determine which initiative is more strategic and will, therefore, result in a greater and more impactful value realization for the company.[21]
8. **Silk Method** quantifies intangible benefits by using seven factors to "harden up" those soft benefits including must-do, faith, logic, direction, size, value, and money. For example, if implementing a system where limited-time value studies have been conducted, this method can prove useful.[22]
9. **PENG Model or (Using Benefits for Prioritizing)** is a tool to identify and assess financial and future or already-achieved benefits. This model aids in communication between business and IT leaders. For example, if the business is working on making operational-excellence improvements and some of those specifications are vague, this model can add clarity and specificity to help manage commitments.[23]
10. **Peer Review** is a model to help with ex-ante evaluations of initiatives and typically involves a committee, cohort, or panel of experts that provide direction to the team to evaluate the quality, credibility, and value of an initiative. For example, before a project begins requirements, an architectural review board may

revalidate team architectural and scale assumptions. This method isn't particularly useful in making prioritization decisions but rather "go" or "no-go" decisions.[24]

11. **Case Studies** is a model to determine value after the fact. Often this method is performed similar to a lesson learned or retrospective for the project team to learn what went right and what could be improved in future projects—e.g., a project went into production, but is this project adding business value?
12. **Strategic Fit** is a method to evaluate and ensure strategic alignment—built on Porter's value chain—between IT investment and the company goals. For example, a portfolio leader has a list of 25 programs and is working to determine which projects have the most strategic influence if delivered this year.
13. **Information Economics** (**IE**) is a model that factors in value and risk over costs and benefits and evaluates people issues in addition to technical ones. This model also can help decide how to allocate resources to be of the greatest organizational benefit. One example of how this model can be applied: If a leader calculates savings for manpower reductions (reducing the labor force), this model would focus on the impact of demotivating staff and changing roles.
14. **Portfolio Management** is a combination of art and science to identify investments that will achieve the long-term company objectives while being mindful of risks. For example, each of a business's 10 leaders proposes 15 projects. However, resource capability can't support 150 active initiatives. Portfolio management can help prioritize and analyze those investments to determine those of most value to the organization based on scope, timing, and expected value delivered.
15. **Internal Rate of Return** (**IRR**) is used as a measure of investments expected and future rate of return—e.g.,

multiple projects have varying delivery dates and value realization, so this model helps to establish common footing for all projects considering the time value of money.

16. **Modified Internal Rate of Return** (**MIRR**) is a model for financially measuring the attractiveness of an investment. It's primarily leveraged in capital budgeting to rank alternative investments of equal size, and it attempts to address some known issues with IRR. For example, if two initiatives will each cost $1.5MM to deploy, MIRR can help identify which is a more optimal investment.
17. **Net Present Value** (**NPV**) is a common financial measure to evaluate a series of cash flows occurring at different times. NPV is best used when comparing similar lives or sizes of projects and can help with project selection, as risk can be calculated when comparing investments. For example, three similar projects are in flight at a company, and funding only allows two to continue.
18. **Equivalent Annuity** (**EA**) is used in capital budgeting to compare mutually exclusive projects with unequal lives by calculating constant cash flow generated by a project over its lifespan if it was an annuity. For example, if a company is considering buying two pieces of equipment, and one has payouts over six years and the other over four years, this model helps determine which is more financially advantageous to invest in.
19. **Payback Period** (**PP**) is a method that assumes a shorter period to repay the initial investment is more beneficial than a longer period, but this method fails to take into account the time value of money or that many IT initiatives have longer payback periods—e.g., deciding between two similar capital equipment purchases that will serve customers in less time.

20. **Discounted Payback Period (DPP)** is a method similar to the payback period that takes into account the time value of money, where the evaluation of profitability and future cash inflows of investment are essential—e.g., two projects have similar payback periods, but project A has 50% of the cash inflows in the first two years, whereas project B has 80% of the cash inflows in years three and four.
21. **Cost-Benefit Analysis Method (CBAM)/Benefit-to-Cost Ratio (BTC)** is a method to compare total costs to benefits, typically using a monetary unit—e.g., if the team invested $100k in a project and, six months later, earned $100k in new sales, the cost-benefit ratio would be one, indicating a one-year payback period.
22. **Economic Value Added (EVA)** is the measure of a company's performance based on the residual wealth calculated by deducting the cost of capital from its operating profit, adjusted for taxes on a cash basis. It's often used for removing conflicting, confusing goals by leveraging a single metric—e.g., a new innovation center is created and, one year later, executives are interested in the economic impact the innovation center has had on company profit.[25]
23. **Accounting Rate of Return**, also referred to as the average rate of return (ARR), is a financial ratio used in capital budgeting—e.g., if a capital purchase is projected to have an ARR of 7%, this means the project is expected to earn seven cents out of each dollar invested (yearly) into the project.
24. **Weighted Scoring Methods** is a model for prioritization of work in which work is scored based on a common set of criteria. Often, this is cost-versus-benefits and is ranked by the final score—e.g., you have two critical projects and use complexity, cost, and time to delivery as primary weighting factors when making the funding decision.

25. **Value of One (N of 1 or V of 1)** is a model that concentrates only on a single observation and then extrapolates for a population, distribution, or sample base—e.g., process-improvement initiatives have resulted in sales managers no longer spending two hours per week on unnecessary paperwork. With a total of 20 sales managers, this translates to 40 hours per week of wasted work.
26. **Before-and-After Model (BAM)** is a model that illustrates the state of the union before an initiative and its state post-initiative. This is a useful method when the difference is subtle between the two or when comparable metrics aren't available—e.g., you joined a team mid-project, and only conjectures were available about the "before" state, but you now have metrics post-initiative.
27. **Operations Performance or KPIs** are key measures that help to define success and track the progress of a team, department, or organization in meeting its strategic goals—e.g., average cost per hospital stay or profit margin by division.[26]
28. **Supplier Business Stability** is a model for assessing business partners to ensure longevity, business alignment, and impact to your bottom line—e.g., the average tenure of consultants on-site.
29. **Value Quantification Framework®** (VQF) is a model that is an organizational methodology for capturing value. This model is best applied to strategic indicatives—e.g., a new process workflow is being introduced and the team believe it will be beneficial but area unclear on specifically which how to quantify the impact of the organizational change initiative.
30. **Critical Success Factors** (CSF) help to explain the key performance areas of an organization. These success factors would be temporary or ongoing, internal or external, building or monitoring, strategic or tactical,

and perceived or actual. For example, executive leadership commitment and experienced staff could both be critical success factors for a project, team, or organization.[27]

31. **Real Options Analysis (ROA)** is an approach for demonstrating valid opportunities that, if invested in today, would result in future benefits. This is often applied when attempting to quantify complex enterprise architecture investments. There are multiple methods for and approaches to calculating real options including partial differential equations (analytics approximations, numerical methods), simulations (Monte Carlo), lattices (binomial, trinomial, quadrinomial, multinomial), close-form solutions (Black-Scholes), and integrals (probability-density functions, analytics approximations, and numerical methods).[28],[29]
32. **Expected Monetary Value (EMV)** is a method used when stakeholders don't agree on the value of an initiative. This model is calculated by taking the probability of achieving a KPI value, multiplying it by the value of a single KPI unit, and multiplying that by the number of KPI units. For example, you attempt to quantify the cost avoided by the introduction of a technology system by getting stakeholders to agree to a 5% savings each time the system is utilized by a single user.

Moving from a Mindset of Project Delivery to Value Delivered

Value management, when implemented properly, is a significant organizational capability that, over time, can become part of an institution's culture.

It's difficult to separate the project-management office from how value is stimulated, surfaced, and shaped. Leaders find it challenging to determine when they don't have the right mix within the PMO to lead transformational change focused on value-based results.

If the topic of value is raised during a meeting, here are three questions to ask, the answers to which may result in you needing to consider separating your strategic initiatives from your traditional project-management office:

1. "What was the value delivered as a result of this completed project?"
2. "How did the project close if the team didn't address the root business problem?"
3. "Where can I go to view the value realized within the last six months for this department?"

Each of these questions begins to uncover a gap between what was requested and what was delivered. Project management is about delivery and, primarily, removing obstacles through risk management. Value management isn't the same thing. Value management focuses on how to realize value by connecting stakeholders and translating their vision so they perceive the worth of something.[30]

These considerations support three mindset shifts that progressive leaders are starting to champion:

1. Project delivery to value delivery
2. Operational improvements to operational excellence
3. Value creation to continuous value delivery

Value management leads the mindset shift from project delivery to value delivered. This changes the conversation from. "How many projects have you completed for me?" to "What was the impact of the project that was completed?" While it may seem like a minor detail, it's actually a transformative change. This mindset moves resources spread out over 100 initiatives to centrally focus on the 50 most critical organizational initiatives and empowers the teams with the proper resource staffing.

Operational improvements slightly but incrementally move the needle. The challenge is, they're not noticeable. The team can identify those changes and likely agree that something changed for the better, but, unfortunately, those leaders looking in on the team can't see that anything changed and only notice a black box.

Operational excellence focuses on the quantification of improvement by using Six Sigma as a base for measured change. The adoption of the entire framework—not solely the terminology—requires a culture change. An operational-excellence initiative is best started within a sub-set of a department or an organization. Start small, and identify a specific problem to solve. Articulate the value achieved and move on. Eventually, this momentum is infectious.

Value creation is the focus of 90% of project-management offices. The focus is on the number of projects per project manager or the number of dollars each project manager is managing. These metrics can be reported, but they're misleading.

Project-management offices are incentivized to deliver projects. In turn, this keeps the focus on executing or creating new projects for delivery. But what's the point of putting something into production if we can't keep the systems up? Leaders need to shift from value creation (pumping out projects for the projects' sake) to continuous value delivery. This means identifying projects that have high potential to deliver value. This brings us back to the value-management office, because traditional project-management offices aren't equipped or trained to address this critical strategic need.[31,32]

Operational excellence and continuous value delivery are the focus of the new value-management office.

I believe that executives who initiate a value-management office and put leaders in place that have a demonstrated history of quantifying outcomes with a background in Six Sigma, portfolio management, relationship management, statistics and mathematics, and learning intelligence will finally be able to articulate the value that their team, department, or organization achieves in advancing strategic initiatives.

CHAPTER 2

APPLYING TECHNIQUES TO QUANTIFY OUTCOMES

"There is no real magic to being a good leader. But at the end of every week, you have to spend your time around the things that are really important: setting priorities, measuring outcomes, and rewarding them."

- JEFFREY R. IMMELT

Chapter Objectives

After reading this chapter, readers will be able to:

- Appreciate why quantification of value matters to leaders
- Experience examples of how to measure seemingly intangible initiatives
- Apply four powerful concepts to quantify value to your portfolio today
- Understand the importance of measuring variation
- Explain dozens of methods in which to quantifiable improve a portfolio, taking a data-first approach

This chapter aims to present a practical approach to applying principles of Six Sigma to statistically control portfolio delivery within a program-management office. Portfolio executives are continuously charged with transforming a low-performing team into a high-performance team. Often, the methods of transformation are based on experience, and these imprecise methods frequently produce inconsistent results. The transformation from a low-performing team to a high-performing team can be accelerated by using a quantified methodology to define, measure, analyze, improve, and control desired outcomes. This chapter presents a practitioner approach to maximizing the organizational outputs from an agile program-management office by leveraging statistics and mathematical principles to tighten process variance.

By applying a fit-for-purpose approach to corporate and operational excellence, portfolio executives can dial in and remediate the root causes of portfolio inefficiencies for maximum agility and value realization.

How Six Sigma Can Transform Your Program-management Office

The application of Six Sigma to portfolio delivery has the potential to transform low-performing program-management offices from conventional cost centers to strategic assets that remove waste, decrease delivery variation, and maximize organizational value.[33]

The majority of research published on Six Sigma falls into five categories: (1) the application of techniques, (2) theoretical qualitative and quantitative statistical models, (3) how to implement the framework, (4) overviews of the methodology, or (5) practical examples based on industry. What hasn't been adequately researched is the applicability of Six Sigma to portfolio, program, and project management. Our focus will be on the relevance of statistical models to solve portfolio delivery challenges.

The research will assume a solid understanding of the history of Six Sigma, statistical models and probabilities, tools and techniques commonly applied to Six Sigma, knowledge of distributions and statistical process control, principles of data visualization, correlations, simple linear regression, and control charts. These topics and many others have been covered more thoroughly in previously published research.

Ralph Waldo Emerson is noted for saying, "common sense is genius dressed in its working clothes." Before buying six sigma implementation guides and racing off to transform your organization with metrics, it's useful to ensure the fundamentals are in place before applying more complex models. It's widely said, that getting to 3-sigma is common sense and beyond that requires a strategy.[34]

Conducting a quick-hit assessment of the portfolio can be one the most effective tools portfolio leaders can leverage. The classic "define, measure, analyze, improve, and control" (DMAIC) can be applied or other approaches can be useful, similar to the below.

1. Mission
2. Map
3. Stabilize
4. Standardize
5. Measure
6. Monitoring
7. Manage
8. Improve

This approach offers a quick method to validate foundational elements, e.g., templates are fit-for-purpose, sponsors are still engaged, and project managers are assigned.[35] Until the portfolio delivery process is mapped, applying models won't be useful as the process is not repeatable and therefore most of the data is effectively random. Project management by its nature is mostly the management of one-offs, and it's difficult to conduct comparisons if the data is random, e.g., project managers are untrained, templates aren't consistently used, or terminology isn't standard across the team. Results from your assessment can be presented using primary tools. For example, a radar graph can highlight capability gaps, a fishbone diagram or Ishikawa diagram can identify causes for delays, and the 5 Whys can dig into root causes. Prior to analyzing and collecting project data, take time to determine which projects to kill, keep or delay.

Our research will focus on the four statistical tools that have a diverse range of applicable uses for portfolio executives: (1) Pareto charts, (2) One-Way ANOVA, (3) One-Sample T-Test, and (4) the I-MR chart. Each chart function will be elaborated with an example later in this chapter. However, we'll begin with a general definition of the statistical model and an example of how each model could be applied to program-management offices.

A Pareto chart is a graphical tool that prioritizes a list of variances based on the frequency of occurrence. This chart is easy to interpret, which makes the application of this model useful for individuals not intimately familiar with a specific project or initiative. For example, a Pareto chart could be leveraged to determine the top three causes of project overspending.

The One-Way ANOVA test highlights statistical differences among the means of three or more independent groups. This model could be applied to determine whether time on the job, years in the field, or education level affect whether a project is delivered on time?[36]

The One-Sample T-Test compares the mean to target[37] and is used with continuous data any time you need to compare a sample mean to a specified value. This is useful when you need to make judgments about a process based on a sample output from that process.[38] Using the One-Sample T-Test to compare data gathered before and after process improvements is a way to prove that the mean has shifted.[39] For example, we might know the average time for a business case to be approved in Q1 last year was 24 days, and we wish to compare this value to the average time for a business case to be approved in Q1 of this year.[40]

The I-MR-chart is a chart of standardized individual observations (I) and moving ranges (MR) from a process. The combination of these two charts allows the observer to track both the process level and the process-level variation at the same time. This chart is handy when trying to view variance over time for a single quality characteristic. For example, if we had wild swings in the portfolio's financial variance—quarter-over—this chart could highlight variances beyond control limits.[41,42]

As we review each statistical model and apply it in a business context, for consistency of delivery, we'll explain each model with the following five elements (an abridged scientific method, for simplicity):

1. Opportunity: the question we're asking
2. Description: the intent of the statistical model
3. Data: a brief explanation of data utilized in the model
4. Charts: creation and generation of visuals for interpretation
5. Interpretation: analysis of results

This format will ensure the application of the models can be consumed by a wide range of readers, observers, and leaders.

Pareto Chart – Variance of Frequency

Pareto charts are one of the greatest tools in the arsenal of an astute portfolio leader. This statistical model allows observers to identify chronic problems within an environment.

OPPORTUNITY

The business case is that projects are continually being delayed for various reasons. Our opportunity or need is to identify the primary causes and develop plans to remediate and control these variations.

DESCRIPTION

The core concept of Pareto charts is that a more significant percentage of problems stem from a lower rate of defects. Applied to our business case, this means that two or three primary drivers are likely causing eighty percent of the project delays. Therefore, instead of boiling the ocean in attempting to solve all problems, we're going to focus on the two or three that will have a more significant effect on getting our projects delivered on time.

DATA

The data set is representative of a list of projects that were delivered in the last six months or are currently delayed within the last twelve months. Initially, project leaders were focused on resources not being adequately assigned to the projects. Our chart surveyed project managers and had the project managers identify the primary reason they felt that the project was delayed.

CHARTS

This chart models the seven primary reasons for delay, in order, by frequency of occurrence.

The seven reasons for delay include:

1. Contract delivery gaps
2. Missed requirements
3. Business unavailable
4. Design gap
5. Resources not available
6. Development delays

7. Vendor misses

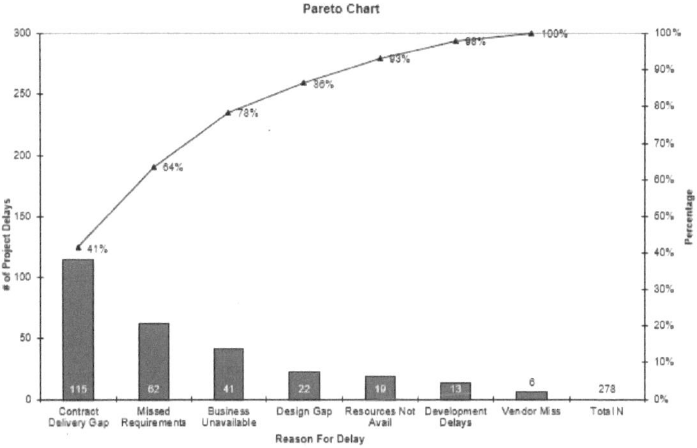

The chart displays the reasons for the delay in descending order of occurrence. The right axis is the cumulative percentage. The x-axis highlights the reason for the delay. The y-axis represents the number of projects delayed.[43]

INTERPRETATION

Initially, we thought that resource allocation gaps might be a driver behind project delays. If we'd raced off to tackle that cause, we would have missed the real root causes. The top three root causes driving project delays are contract delivery gaps, missed requirements, and business not being available.

It's probable, based on experience, that the business not fully participating resulted in missed requirements. These were likely captured relatively soon, however, after the contract was signed. The effect is that the vendor is pressured to deliver but isn't contractually bound to provide what we, the client, require.

The chart is powerful and focuses the project team on occurrences that will have the most significant effect on outcomes.

One-Sample T-Test – Compare Mean to Value

The One-Sample T-Test compares the mean of one distinct group to the mean of another group. This statistical test compares the mean of a single sample to a predetermined value to assess whether the same mean is significantly greater or less than that value. This statistical model is excellent at comparing two matched scores or measurements such as before vs. after, e.g., comparing a sample with a benchmark.[44]

One requirement for this test is that the population is known and, therefore, we can estimate the population standard deviation using sample data.

OPPORTUNITY

The business case is that we conducted an assessment of our project managers in quarter one. We need to determine if the changes we implemented—e.g., training, mentoring, a new onboarding process—was statistically significant and had an effect on performance.

DESCRIPTION

The One-Sample T-Test is used to estimate the population mean from a sample and, in this case, perform a hypothesis test for a mean. Using a One-Sample T-Test to test the difference of the same subjects at two different times is commonly called a Paired T-Test. We'll also be using a two-tail test, because we could have performance scores that either increase or decrease. Also, because we've tracked these project managers over two quarters and they're the same individuals, we'll be applying a Paired T-Test.

The One-Sample T-Test will help us compare a test variable against a known test population mean to understand how changes made may influence future performance outcomes.

DATA

The data set is the result of an assessment of project managers in the first quarter and the work of a second assessment conducted during the second quarter.

Each project manager was evaluated twice—once during the first quarter assessment and also during the second quarter assessment.

CHARTS

The result of conducting this One-Sample T-Test is below. This is the outcome from our dataset spanning two quarters.[45]

SUMMARY			Hyp Mean	3.09
Groups	*Count*	*Mean*	*Variance*	*Cohen d*
Quarter 1	33	3.093939	0.612462	
Quarter 2	33	3.518182	0.491534	
Pooled			0.551998	4.730019

T TEST: Equal Variances				Alpha	0.05				
	std err	*t-stat*	*df*	*p-value*	*t-crit*	*lower*	*upper*	*sig*	*effect r*
One Tail	0.182906	19.21343	64	1.43E-28	1.669013			yes	0.923172
Two Tail	0.182906	19.21343	64	2.86E-28	1.99773	-0.78964	-0.05885	yes	0.923172

T TEST: Unequal Variances				Alpha	0.05				
	std err	*t-stat*	*df*	*p-value*	*t-crit*	*lower*	*upper*	*sig*	*effect r*
One Tail	0.182906	19.21343	63.24121	0	1.669307			yes	0.923982
Two Tail	0.182906	19.21343	63.24121	0	1.998191	-0.78972	-0.05876	yes	0.923982

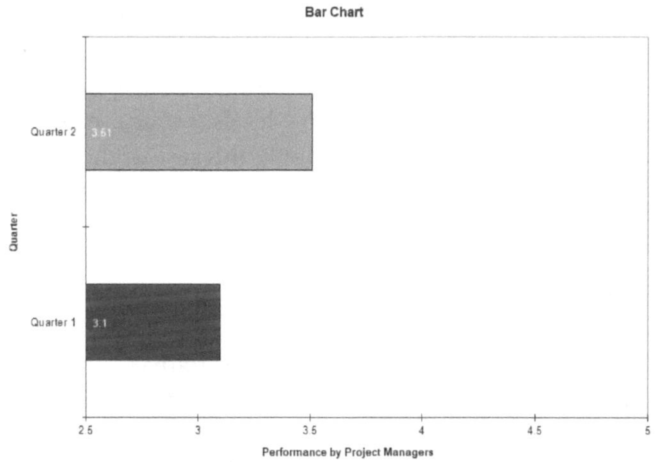

To make things a bit clearer, we charted the mean performance-assessment scores by the project manager for the first and second quarters. This visualization allows us to see if we have a difference in the scores.[46]

While it's evident there was likely a change, we didn't know what direction that change would be in. As a result, by generating a simple bar chart, it makes visualizing the difference between quarter-one and quarter-two assessments easier to interpret.

INTERPRETATION

After analyzing our results, we can observe that the p-value = 1.43E-28 and is very low. We interpret this as having a very low probability that our data is random. Our confidence is extremely high that there's a statistical difference between quarter one and quarter two; the changes we implemented have had a very positive effect on project-manager performance results.

One-Way ANOVA – Differences Between Groups

One-way analysis of variance of ANOVA is used to compare means of two or more samples. A One-Way ANOVA test uses a single independent variable. There's also a Two-Way ANOVA test that uses two independent variables. The independent variable should have at least three levels, groups, or categories. In our business case, we'll apply this test to three different vendors. However, a One-Way ANOVA can be used with two levels, in which case it's equivalent to a t-test.[47]

OPPORTUNITY

The business case is that we have three vendors we're sourcing project managers from for current initiatives. One of the vendor partners insists that they have a solid track record of providing high-quality project managers. In fact, after speaking to all three vendors, none indicate they offer anything except exceptional performance. When we push, managers acknowledge specific individuals were terrible hires but identify those individuals as exceptions.[48]

Which vendors should we slow hiring from, and which vendor should we increase hiring from? We have three hundred and forty-nine project managers engaged across all three vendors because we don't have time or resources to interview and assess all project managers. We polled a sampling of ten project managers from each vendor. We'll use the ANOVA test to extrapolate our results.

DESCRIPTION

The ANOVA test, or analysis of variance, tests variation of two or more independent groups to determine if there's statistical evidence that the target-population means are significantly different.[49,50]

DATA

The data set was based on a performance review conducted on ten project managers sourced from three different vendors. There was a suggestion that at least one vendor is underperforming the other vendors. We've conducted a performance assessment on ten project managers from each vendor as a sample population. We've leveraged that data to conduct the following analysis.

CHARTS

This analysis determined there are statistically significant differences between the two groups.

ANOVA: Single Factor

DESCRIPTION					Alpha	0.05		
Group	*Count*	*Sum*	*Mean*	*Variance*	*SS*	*Std Err*	*Lower*	*Upper*
Vendor 1	10	36	3.6	1.155556	10.4	0.23737	3.112957	4.087043
Vendor 2	10	35.6	3.56	0.122667	1.104	0.23737	3.072957	4.047043
Vendor 3	10	26.1	2.61	0.412111	3.709	0.23737	2.122957	3.097043

ANOVA

Sources	*SS*	*df*	*MS*	*F*	*P value*	*Eta-sq*	*RMSSE*	*Omega Sq*
Between Groups	6.280667	2	3.140333	5.573457	0.009413	0.29221	0.746556	0.233656
Within Groups	15.213	27	0.563444					
Total	21.49367	29	0.741161					

Similar to the analysis in the above chart and to make things a bit clearer, we charted the mean performance scores by the vendor in a simple bar chart for better visualization.

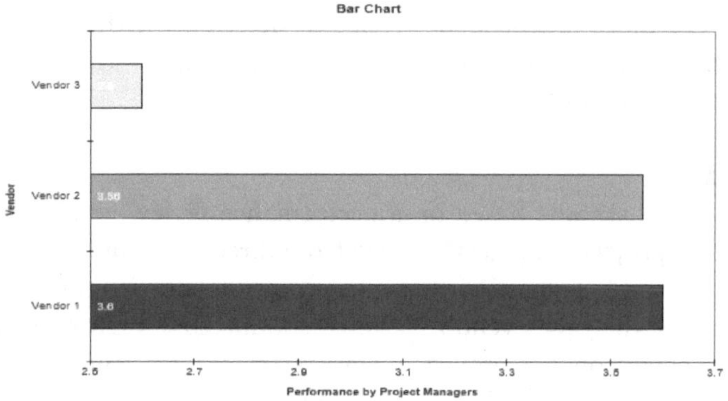

We now can see apparent differences in how vendors are performing when we observe the mean vendor project-manager scores with n = 10 per vendor.[51]

INTERPRETATION

A p-value of greater than .05 suggests that there were no statistically significant differences among the group means as determined by the One-Way ANOVA. In this case, reporting the standard deviations can still be useful.[52]

A p-value of less than .05 suggests there were statistically significant differences among the group means as determined by the One-Way ANOVA. In this case, you may want to run a post hoc test. Remember, the ANOVA test tells you whether you have overall differences among your groups, but it doesn't tell you which specific groups are deferred. This is the value of conducting a post hoc test. We aren't going to cover post hoc tests within this research in depth. Various resources do cover basic tests in greater detail.

As a reference, if your data met the assumption of homogeneity of variances, use Tukey's honestly significant difference (HSD) post hoc test. If your data didn't pass the homogeneity of variances assumption, consider running the Games Howell post hoc test.

In our example, we have a <1% p-value or significance level. This indicates we have a <1% risk of concluding that a difference exists when there's no actual difference. As a result, we can determine there was a statistically significant difference between group means as determined by One-Way ANOVA.

Analyzing the bar chart, it's quite evident that Vendor 3 is performing considerably below Vendor 1 and Vendor 2. This means that the average project managers from Vendor 3 are of substantially lower quality than the project managers from the other vendors. As a result, this will impact our future hiring practices and how we select talent.

The changes we introduced—to push change—did, in fact, have a statistically significant effect.

I-MR-Chart – Observation Variance Against Limits

Control charts measure and can be used to monitor process characteristics over time. These charts allow for a broader perspective and can more easily gauge changes in a variable over time.

I-MR charts visually display standardized individual observations (I and moving ranges (MR) over time. Individual observations are charted above the graph for moving ranges. The benefit of seeing both charts is you can view your process level and process variation at the same time.[53]

OPPORTUNITY

The business case is that we have a recurring issue of managers overspending on purchase orders. A project manager drafts a contract, and that contract is submitted to procurement by a company manager. There are recurring incidents where vendors have submitted invoices for work that's been completed but hasn't been approved through procurement. As a result, this spend is unauthorized and a violation of company policy. Nonetheless, the liability of work performed has already been incurred, putting the company in an at-risk position of owing money for work earned without a contract.

DESCRIPTION

The I-MR chart is a control chart for monitoring the mean and variance of a process that's subject to shifts and drifts. By recording sample observations of the process, we can determine slow drifts in performance—in this case, before the year end is reached.[54]

DATA

Our data represents the number of overspending incidents during the quarter. As an example, a value of four indicates that, four times during the quarter, spending over the purchase-order limit occurred. This data was recorded quarterly, and all incidents that occurred during the quarter were reflected in that data point.

CHARTS

The top graph, or individual chart, tracks the values of each observation. The bottom graph, or moving-range chart, plots process variation as calculated from the ranges of two or more successive observations.

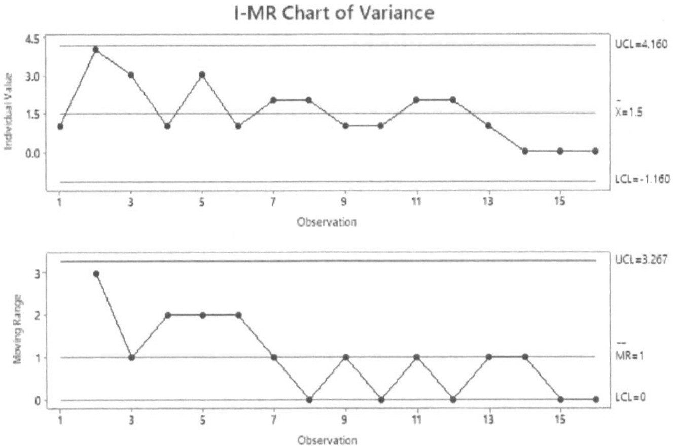

Often, a simple control chart is excellent for tracking financial variance to upper and lower boundaries. In this case, because of the data set, it's slightly easier to visualize this data and the reduction of variation with an internal plot for the variance.[55]

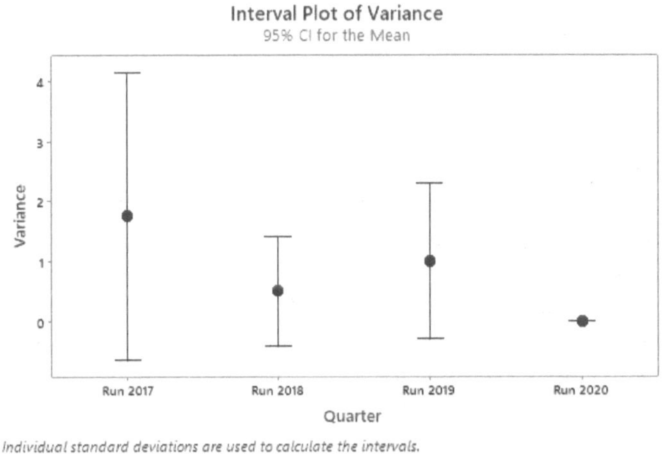

Using both graphs, we can now readily determine if our changes have resulted in a favorable outcome.

INTERPRETATION

In the I-MR chart of variance, we can view the incidences of purchase-order overspends charted beginning in quarter one of 2016 through December 2019.

The mean value is represented by the green (center) line. The upper and lower control limits (above and below) are represented by red lines. A quarter (or process) that's out of control might show variation due to unique causes. First, we look at individual values. If no red points outside the upper control limit exist, the process is in control. Next, observe the moving range or mean and determine whether the process mean is in control using the same process to interpret the earlier chart. We're looking for a red outliers. If the moving range is stable, we can now interpret this chart as "no failed observations."

We can observe visually that the individual values started as a worse case of four incidents per quarter in 2016, three incidents in 2017, two incidents in 2018, and zero incidents in 2019. The changes that were made to narrow process variances were effective.

Research Propositions

There are myriad applications where Lean Six Sigma and Six Sigma principles can be applied to quantifiable delivery outcomes for use by executives accountable for large portfolios. It's equally possible to believe that leaders must be quants, mathematical experts, or have in-depth statistical process knowledge to demonstrate portfolio impact. Of course, a background of data science helps. However, how do you train and provide knowledge transfer to your team? We must simplify complex analytical models into knowledge and intuitive visual insights.[56]

Four propositions (P1, P2, P3 and P4) are offered that have the potential to introduce impactful methods for transformation into a portfolio-management team that's ripe for a win.[57]

PROPOSITION 1: DEPARTMENTAL PROJECT MANAGEMENT MATURITY TRACKED WITH RADAR CHARTS

As new leaders join and are charged with leading existing teams, one of the first questions asked by executive leadership is, "What impact have you had on outcomes? How are you moving the needle? Can you quantify the highlights of the team over the past, month, year, or quarter?"

To perform this analysis, the leader needs to have assessed the team initially. This critical step is often forgotten. The leader steps in, provides guidance, restructures the group, and offers insights. However, they didn't take the time to capture the current state.

Define the current state. Identify the desired future state. Quantify that gap. One approach, simpler than using a One-Sample t-test, can be to measure sustainable improvements using a radar chart illustrating the core capability areas that you're interested in measuring and maturing.

PROPOSITION 2: THROUGHPUT ANALYSIS OF CONTRACT EXECUTION WITH BOX-AND-WHISKER CHART OR BOX PLOT

When a process breaks down, it's natural for team members to share the concern that it's not within their span of ownership to make a change to affect positive change. It's true that if you don't own a step, and that step is slowing down throughput, there might not be much you can do. However, no one wants to be on the team that's slowing down the process.

If the contract duration excessively lengthens and it is further complexified by multiple teams involved in delivery (finance, procurement, legal, etc.), modeling the handoffs with a box-and-whisker chart can be instrumental. The result is a common-sense visual identifying the before-and-after effects of process change.

PROPOSITION 3: CRITICAL PATH ANALYSIS LEVERAGING THE LAST PLANNER SYSTEM

The Last Planner ® System (LPS) is a principle largely applied to construction to promote collaboration between tradesmen. The goal is to increase trust within teams by identifying future work (six months in the future) and taking steps today to collaborative with parties required to bring those dates in either days or weeks. Innovative portfolio leaders are applying this concept to promote collaboration with cross-functional teams, resulting in shorter delivery timelines.[58]

This approach has four levels of planning: (1) master scheduling (setting milestones), (2) phase scheduling (specifying the handoffs, collaborative planning, and reverse phase scheduling), (3) lookahead planning (breaking down tasks, identify and removing constraints, designing operations), and (4) commitment or weekly work plans (weekly work assignments, reliable promising, action reasons for delays, learning). This process can also be described as should, can, will, and did planning.

By looking ahead toward the next three or six months, project resources and collaboratively pull-in the schedule. If we can push the schedule out, why can't we bring it in? This approach can be extremely impactful for rebuilding trust and improving delivery of teams assigned across disciplines.

PROPOSITION 4: PORTFOLIO COMPLEXITY ANALYSIS WITH PROJECT COMPLEXITY ASSESSMENT AND MANAGEMENT (PCAM)

It's enviable that success breeds success. The more effective your program and project management delivery teams are, the more initiatives they'll be asked to champion and get over the finish line.[59]

A project-complexity assessment-and-management (PCAM) tool can help to identify the critical project complexity factors and assess the potential of each to impact the outcome of the project. This can be a great tool to deploy at multiple stages of a project (iterations or sprints) in various ways to evaluate the risk of development and the risk of complexity on the ultimate project delivery. The PCAM tool can help teams identify, assess, and manage project complexity.

Limitations and Direction of Future Work

This chapter contributes to the limited body of knowledge directly attributed to Six Sigma applications within the context of program-management offices. The application of statistics and applied mathematics heavily depends on data quality. While I believe in the application of Six Sigma to achieve quantifiable, program-management office improvements, the identification of the most appropriate statistical model to analyze variance is critical. As a result, it's possible to have ninety percent of the right intent with a statistical model that isn't the most applicable model for the data set being analyzed.[60]

While I stand behind the vision that statistical process control can be almost uniformly applied to the program-management office to quantifiably measure continuous improvements, I acknowledge there are limitations.

One additional important concept is that all data is not standard or consistent and therefore it's not always possible to combine data for collective insights. Clearly combining data for better insights is a desirable objective. However, in the case of multiple, teams, departments or divisions how data is collected or which data is collected might vary. The impact on our ability to make assure quantified analysis is that some data we don't be able to compare to each other. Often the lack of consistent data is not intentional but rather an outcome of non-standard processes.

With that context, below is a list of additional statistical models that can be applied to accelerate the delivery efficiencies of a program-management office. The following statistical models could be explored in greater depth and applicability to quantifiably improve potential within portfolio, program, and project-management functions and capabilities.

1. **Two-Sample T-Test:** Determine if the average of one group of data is more significant than (or less than) the average of another group of data; e.g., compare the project-management maturity between two project-management teams.
2. **Regression:** Look for trends, outliers, or oscillations against a factor, e.g., are more projects delayed in December than March?
3. **Box-and-Whisker or Box Plot:** Compare median and variance between two groups of data or to identify outliers; e.g., quantify how much contract execution has slowed between two time periods.
4. **Cause-and-Effect Diagram/Fishbone:** Brainstorm the possible sources of variance for a particular event; e.g., explore why a critical project risk wasn't identified earlier that negatively impacted project delivery.
5. **Two-Way ANOVA:** Compares the mean differences between groups that have been split into two independent variables or factors. This helps determine if there's an interaction between the two independent variances on the dependent variable; e.g., to determine if the team's performance has improved six months into a new manager's assignment to the unit.[61]
6. **Z-MR Chart:** A stacked chart with the top chart plotting individual observations and the bottom chart plotting the range of individual observations; e.g., quarterly financial forecasts are inaccurate, and controls were put in place to narrow variance. This model could help visualize if those changes are statistically significant and, in fact, having a positive effect on financial accuracy.[62]
7. **Design of Experiments (DoE):** Systematic process to break down the variables of production and analyze each one using the factor, level, run, response, and application; e.g., an integrated project plan was created for a complex multi-year project and you want to un-

derstand which resources to add (a manager, an engineer, a developer). Enter the design of experiments.

8. **Takt Time**: Maximum acceptable time to meet demand or the process duration matched to demand; e.g., contracts consistently are reviewed that have been impacting delivery. You want to determine what KPI is required to meet your departmental demand or contracts reviewed per week. In leading an agile team, you want to understand the mean time for story completion; e.g., the team completes 1.63 stories a week. This offers a more predictive forecasting approach when compared to story points or burn down charts.[63]
9. **Monte Carlo Method**: This is a computerized, mathematical technique that allows people to account for risk in quantitative analysis and decision-making; e.g., financial estimates vary widely for your department. You apply a distribution model and determine the probability you'd hit your quarterly budget based on past performance.
10. **Binomial Probability**: the probability of exactly x successes on n repeated trials in an experiment which has two possible outcomes; e.g., new projects are kicking off and concerns exist about potential future project delays. You evaluate the last sixty days of project delivery and determine the percent of projects, that started this week, which will be late.[64]
11. **Cost of Poor Quality (COPQ)**: The cost that would disappear if the systems, process, and products were perfect; e.g., your ServiceNow instance is eating tickets. You want to build a business case on the cost avoided if adequate resources were assigned to manage and triage tickets.
12. **Failure Mode and Effect Analysis (FEMA)**: A method of analyzing the ways or modes in which something might fail; e.g., as a product owner, you launch a new

product feature but support costs are unusually high. What's driving those defects?[65]

13. **Paired T-Test:** Used to compare two population means where you have two samples in which observations in one sample can be paired with observations in another sample; e.g., a new product was launched, and after training, you were asked how practical your training module was at improving team knowledge retention.
14. **One-Sample Proportion Test:** Used to estimate the proportion of a population. It compares the proportion to a target or reference value. Also, it calculates a range of values that's likely to include the population proportion; e.g., a new PMO process was implemented that takes fifty percent less time to execute. You recently closed ten projects and want to know if the process has improved beyond the fifty percent previously baselined.
15. **Multiple Linear Regression:** Used to model the relationship between a continuous response variable and continuous or categorical explanatory variables; e.g., you're frequently assigned PMs from a vendor with various work experience, certifications, and industry experience. With a recent resource swap, you want to calculate the average monthly delay of swapping out a resource with fifteen years' experience for one with five years' experience on project delivery.[66]
16. **X-Bar Chart:** Used to monitor the arithmetic means of successive samples of constant size, n. This type of control chart is used for characteristics that can be measured on a continuous scale, such as weight, temperature, or thickness; e.g., you want to determine if the average delivery per PM varies over the course of a year.[67]
17. **Histogram:** a graphical display of data using bars of different heights to display the shape and spread of continuous sample data; e.g., sampling projects over

the last three months to visually show 95% confidence of starting on time. You want to show, of the twenty projects, in our queue with dates, which projects will start on time to better manage sponsor expectations.

18. **R-Chart:** Used to monitor the variation of a process based on samples taken from the process at given times (hours, shifts, days, weeks, months, etc.); e.g., an agile team is entering Jira tickets into a system to represent demand. You want to determine if the pace of tickets is steady over time or varies.
19. **U-Chart:** An attributes control chart used with data collected in subgroups of varying sizes; e.g., your vendor's agile team completes stories each week, and you want to view a trend of successfully completed stories to determine if the process the team is leveraging is stable and predictable.
20. **P-Chart:** Used in statistical quality control to graph proportions of defective items; e.g., it's the final week of testing, and you want to track the pass or fail events for test cases (regardless of the number of defects). You highlight the number of tracked failures over total units tested.[68]
21. **5 Whys:** The risk to citizen, government, and business data loss resulting from seemingly ubiquitous access to services. An exploratory technique to use an interrogative approach to determine the underlying relationship between cause and effect; e.g., if a vendor deliverable was missed, we incorrectly assume they had sole accountability when, in fact, the root cause, discovered by the 5-Why exercise, was that we provided requirements to the vendor three weeks late.

This proposal presents various statistical methods to quantifiably improve portfolio outcomes using Six Sigma principles. However, further research is required in three areas:[69]

1. **Critical Path Analysis:** Virtually every delivery has a critical path for success. The application of Six Sigma to evaluate the risk of critical-path delivery is a great area of untapped potential.
2. **Education:** The tools to transform portfolios are almost limitless. Unfortunately, there's very little knowledge or awareness of the applicability of these tools from a practitioner perspective. Additionally, education and training could make transformative gains in this space toward the wide adoption of Six Sigma principles.
3. **Vendor Risk Assessment:** Companies are shifting noncore capabilities to vendor partners outside the walls of the organization. Initially, this appears to be a great idea; however, back sourcing quickly occurs, increasing the burden on internal employees when external partners fail to deliver consistently. Six Sigma has great potential to quantify vendor impacts before they occur.

For example, on a multi-year, new, critical pharmaceutical launch, applying Six Sigma tools could be a game-changer. This requires a mindset shift from hiring existing skills the organization likely already has in house to onboarding new resources that have a combination of portfolio delivery, Six Sigma, and relationship-management skills to drive change organizationally.

These additional challenges require practical solutions and further research before we'll be able to raise the level of Six Sigma intelligence uniformly across the organization.

The Impact of Quantified Outcomes

Six Sigma can be applied to quantify outcomes in portfolio delivery. Many progressive leaders have already started to adopt pieces of Lean Six Sigma, operational excellence, continuous improvement, and continuous delivery in their quest to achieve 10x results.[70]

If you raise the topic of Six Sigma in an executive discussion, you'll likely hear the following:

1. "We need to focus on the work at hand. We don't want to be in analysis paralysis."
2. "If people don't understand the statistics, change won't happen."
3. "It's interesting; let's review that in a few months."

Ultimately, change isn't about statistics or mathematical computations or data integrity. Change is about people. Similarly, Six Sigma isn't about quantifying this or analyzing that. Six Sigma, when applied to portfolio excellence, is about achieving sustainable change through people.[71]

To have continuous impactful results, companies must have three supporting structures in place. First, the CEO, president, or COO must acknowledge that Six Sigma allows them to make changes faster and more efficiently.[72] Second, a vice president or Six Sigma champion must be selected to provide the necessary support and infrastructure for the role. The third supporting structure is the black belt or master black belt. An employee with deep delivery execution, mathematics, statistics, and a computer-science background must be empowered to lead change initiatives.

Six Sigma is about macro and micro adjustments. The macro adjustments are visible. For example, you're introducing new methodologies for analyzing outputs or rolling out a new framework for capturing value. However, micro adjustments are equally important. If you're preparing to fly on a plane cross country, you see the airplane door close and hear the flight attendant's call to buckle up. Those are macro changes. The micro changes are less visible but nonetheless critical—for example, the pilot makes a decision to deviate five degrees to avoid weather that would result in unnecessary turbulence for passengers, and then you receive the update that you'll be arriving at your destination airport twenty minutes early. That micro decision resulted in saving twenty minutes during that flight.

Portfolio executives shoulder a lot of responsibility for unparalleled outcomes—every leader wants more for less out of their team. However, only visionary and progressive leaders have the intuition and confidence to empower Six Sigma portfolio champions to drive transformational change.

With risk, there's a reward. Transformational CEOs, CIOs, and COOs are investing in Six Sigma leaders to quantify change and enlighten the path to exponential results.

I believe portfolio directors who leverage Six Sigma to drive continuous improvement within their portfolios contribute more significantly to organizational value maximization.

CHAPTER 3

A STEP-BY-STEP GUIDE TO ARTICULATING VALUE IN PROJECT MANAGEMENT

"Operations keeps the lights on, strategy provides a light at the end of the tunnel, but project management is the train engine that moves the organization forward."

- JOY GUMZ

Chapter Objectives

After reading this chapter, readers will be able to:

- Identify how to capture value in your initiatives
- Apply quantified approaches to explain outcomes
- Understand how to progress from ideations to a quantified value-based outcome
- Appreciate the differences between current and future state mapping vs. value stream mapping
- Transform a value stream map into quantified financial value

This chapter aims to present a practitioner approach to applying project-management principles of value realization within a value-management office. Portfolio leaders and program and project managers are confused about how to apply value management in the real world. Project-management offices are being shut down, because they've not delivered on the promise to add value to the business. What has emerged to fill this void is the concept of a value-management office. This new organizational construct is narrowly focused on maximizing value for the company's strategic initiatives. The result is a huge divergence from the traditional project-management office, which has been viewed organizationally as a cost center that provides little if any value to the company.

To optimize the value-management office, new techniques need to be applied and integrated into the existing processes and procedures to deliver projects and quantify results. By using a general-structure portfolio, program and project managers can deliver impactful and quantified value to business partners. In this chapter, we'll explore an example of how to achieve value realization. Leaders, by using this model, have a roadmap for achieving quantified outcomes for their value-management office.

How to Demonstrate Value

Value is easy to see when it's in front of you and hard to explain when it's not. Value-management-office leaders have trouble explaining what they mean when they say, "We need value delivered, not projects delivered."

Throughout this chapter, we'll cover specifically what it means to demonstrate and quantify value for a project or initiative. There are lots of great tools and techniques that can be applied when driving a results-oriented project. Pareto charts can help with categorizing variance in the form of defects. Failure-mode and effects analysis can identify process failures. The design of experiments can aid in the interpretation of control tests to evaluate influencing factors. Each of these methods can be very useful. These methods are covered in greater detail in other works, so we won't be covering them here.[73]

We're going to focus on a simple example to demonstrate the process of creating value and how creating quantified results can be straightforward if you follow a repeatable, six-step process.

1. Improvement metrics
2. Process mapping and waste identification
3. Current and future state value-stream mapping
4. Value management and value-stream comparison metrics
5. Lean comparison metrics
6. Financial savings

Improvement metrics begin with the end in mind. These metrics highlight the before-and-after desired results you're expecting. These values can change and evolve throughout the project; however, the intent is that you have something in writing before you start. For example, are you going to decrease cloud storage needs by 20%, increase speed to market by 10 days, or compress a 30-day process down to 10 days? These are potential metrics that you could start with. They're clear, defined, and documented.[74]

Process mapping and waste identification defines a process map of the current and future state. The process map captures the voice of the customer and their interactions. It also identifies and classifies process inputs and outputs including the associated tasks. These models can be extremely complex or consist of a few (usually five to fifteen) keys steps. Waste identification addresses the types of waste for each process and quantifies the degree of waste.[75]

Current- and-future-state value-stream mapping is exactly what it sounds like. The current-state value-stream map defines the current process as-is, unchanged. In contrast to process mapping, value-stream mapping also identifies waste within and between processes, while process mapping only identifies and classifies inputs and outputs. This is an important distinction. Later, we'll use these maps to help quantify improvements.

Value management and value-stream-comparison metrics contrast the current-state value-stream map to the future-state or final-state value-stream map. The result is a clear and quantified picture of the impact the team had in improving the process, project, or initiative.

Lean comparison metrics compare the baseline to the improved process. This is an important step that allows the team to determine the effects of changes to the process on the percentage of value-add and percentage of non-value-add activities within the process.

Lastly, financial savings capture the impact the project can claim in terms of annualized or monthly hard and soft savings. Typically, this process is also referred to as the cost of poor quality (COPQ). Project leaders need to follow their company's policies and procedures on how to evaluate project savings. However, this provides a great start.

This chapter provides a guide for value-management leaders to better understand the process of quantifying value for their team, department, and company. Please use this as a reference and guide for your future projects.

The Starting Point to Clarify Value

Great projects start with a clear problem statement. Beginning with a clear definition of the project and the expected benefits can align stakeholders and simplify future project updates.

What you define in your statement of need is important, but how you define the statement is not. Often, project teams leverage the concept of a project charter to capture the voice of the customer and the stakeholders' expectations for the initiative. There are a ton of templates available in the ether, so I won't be covering examples. I do, however, want to talk about what's often missing—the metrics.

A project charter usually covers the schedule, scope, budget, quality, and benefits. The benefits have some type of high-level, abstract, and hard-to-understand outcomes that can't be measured.

We're going to change that up. We want our improvement metrics to be quantifiable. The project leader working with the team needs to decide on what metrics to lock in. Improvement metrics have five elements: metrics, baseline, target, actual, and percentage change.[76]

The numbers are important. What's equally important is, indirectly, what they say. They address why the project is important and define the real problem to be solved. They capture what success looks like and who's benefiting from the project. The metrics also factor in how long the positive effects of the project will last. This hyper-focus on the problem statement clarifies where improvement efforts are to be focused.

In an effort to get started, project leaders often overlook the most important part of a project—clarifying the problem to be solved. In reality, this process doesn't take long to complete, if the right people are assembled from the beginning. This ensures that the problem statement reflects all critical stakeholders. The table below shows one example of improvement metrics for a project initiative. Actuals are forecasted until the project completes.

Metric	Baseline	Target	Actual	Percent Change
Average contract cycle-time (days)	42	20	10	-76.19%
Average time to socialize contract (days)	23	10	5	-78.26%
Average contracts processed weekly	0.5	3	2	300.00%

TABLE 1.0 IMPROVEMENT METRICS

We've now defined our metrics and have a clear picture of the problem statement we're going to tackle.

Process Mapping and Waste Identification

You can't change what you can't measure. We need a clear picture of what's being performed today before we make changes or suggest improvements. We may want to deploy new data-analytics capabilities, roll out a new enterprise resource-planning module, or even champion a business-process-transformation initiative. This approach works for all of them.

Focus less on *what* you're doing and more on *how* you're doing it. Lean in and try to understand each step and the goals of that step. Ask yourself, "Why is this step important?" If it isn't important, we consider that waste and would have already removed that step for you.

CURRENT-STATE PROCESS MAP

The goal of the current-state process map is to define what's performed today. This represents the as-is or present state of the process or workflow.

Remember, this workflow may involve storing documents on a network, but it could be almost anything. It could represent a continuous-improvement and continuous-delivery pipeline for DevOps. It could represent a technology replacement, swapping a new system for an old one, or trying to use robotic-process automation to save time from human-based activities. This approach can streamline the processes you'll automate to save time before you build the intelligent agents to run jobs. The range of use cases is almost limitless.

Start by selecting a standard modeling technique. There have been many business-process modeling notations that have sprung up over the last 20 years. The Object Management Group now maintains Business Process Modeling Notation (BPMN). The Unified Modeling Language (UML) is another popular, general-purpose modeling language. There's also standard modeling language embedded within tools such as Visio, LucidChart, or Minitab. Regardless, choose a modeling language that works for you.

For our purposes, I'm going to select the simple example of a contract-management process. In this process, the team is modeling the process to execute a statement of work. There are only five core steps in the model, starting with drafting the SOW and ending with reviewing the SOW for submission.

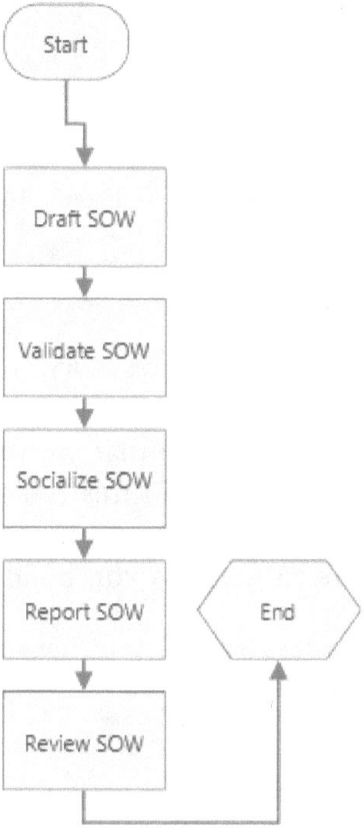

ILLUSTRATION 1.0 CURRENT-STATE PROCESS MAP

For simplicity, I've used the same key steps and altered the cycle-time, wait time, and process time of each step to keep our example straightforward.

WASTE ANALYSIS BY OPERATION

Once the current-state model is drafted, we can review the process map for areas of waste. Essentially, we're looking to identify areas within the model that aren't efficient.

Below is an example of a waste-analysis-by-operation table that highlights where in our process we have waste.

	WASTES								
Operation/Process Step	Overproduction	Waiting	Transportation	Overprocessing	Excess Inventory	Unnecessary Movement	Defects	Waste of Human Capital	Comments
Draft SOW	0	3	0	0	0	0	0	0	None.
Validate SOW	0	6	0	0	0	0	0	0	None.
Socialize SOW	0	9	0	0	0	0	0	7	None.
Report SOW	0	3	0	0	0	0	0	5	None.
Review SOW	0	0	0	0	0	0	0	0	None.

TABLE 1.1 WASTE ANALYSIS BY OPERATION

There are seven areas of waste that we focus on generally. However, as you can observe in this table, not all areas in our example have waste. This list of waste areas can help you narrow down where to focus as you build your future-state process map:

1. Transportation
2. Inventory
3. Motion
4. Waiting
5. Over-processing
6. Overproduction
7. Defects
8. Poor resource utilization

Transportation addresses excess movement of work-in-process, information, or material. Inventory covers information or materials in queues. Motion involves wasted movement made by people while working or performing their job. Waiting includes lost time due to poor workflows. Over-processing deals with work that adds no value to the customer or business partner. Overproduction is the creation of excess supply beyond the requirements of the next process. Defects are the result of re-work when the product or service is out of specification. Poor resource utilization is the ineffective use of energy, people's time, or not fully utilizing people's abilities.

With the problem statement defined, improvement metrics in hand, current- and future-state process maps, and our waste analysis by operation, we can now progress to performing our value stream mapping.[77]

Current-and-Future Value Stream Mapping

The current- and future-state value-stream mapping exercise has three primary components.

First, we define the current value-stream map. This can be referred to as an activity diagram. Second, we identify areas of waste within the current-state value-stream map. When we do this, we tag those areas for Kaizen events. Third, we take the learnings and improvements from the Kaizen events and develop a future-state value-stream map.

I've combined all three of these illustrations into a single page (below) for ease of reference and simplicity.

CURRENT-STATE VALUE STREAM

This is one of the most critical steps. The current-state value-stream map identifies activities or separate entities. The objective is to visualize the process of collecting, viewing, sorting, and organizing data elements into a single view.

Shapes can represent a process, inventory, waiting, information, outside sources, Kanban events, load leveling, and more.

What's important here is that the key process steps are represented. Once those essential steps have been modeled, the next step is to add information to those process steps.

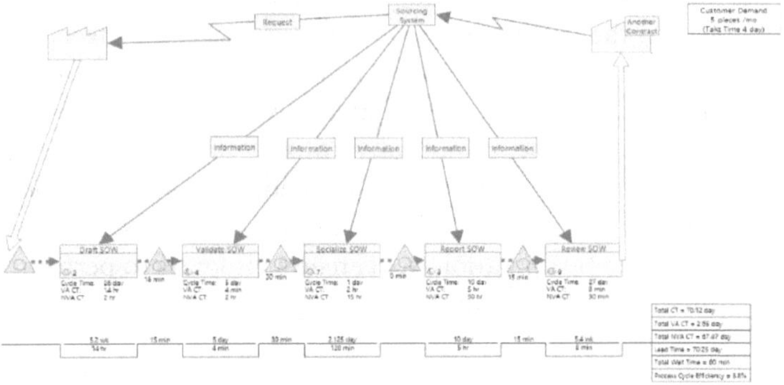

ILLUSTRATION 1.1 CURRENT-STATE VALUE STREAM

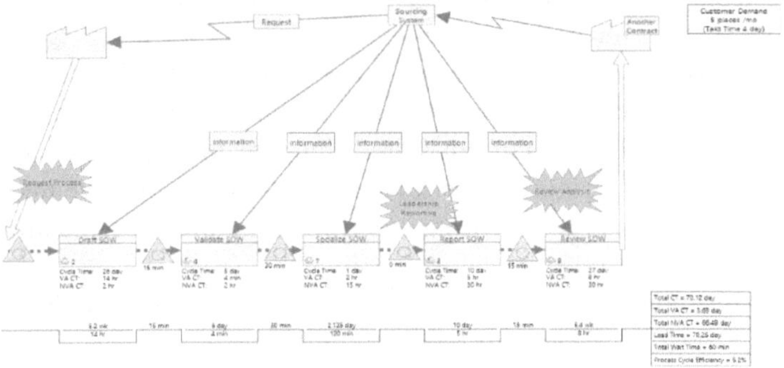

ILLUSTRATION 1.2 KAIZEN EVENTS

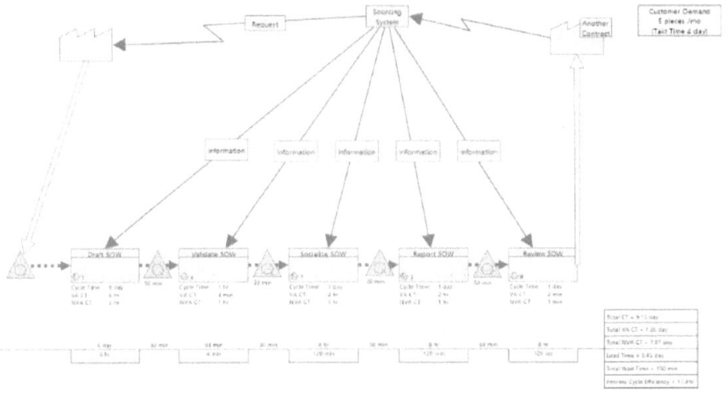

ILLUSTRATION 1.3 FUTURE-STATE VALUE STREAM

There may be a lot of information to collect. However, use your judgment on what's essential for the project, initiative, or transformation you're accountable to deliver. There are generally several major areas of data to leverage:

1. Mapping
2. Processing
3. Inventory

Let's cover the major fields you'd likely need to collect to generate intelligent quantification-of-value reports.

Mapping

- Customer demand
- Shift information
- Takt time

Processing

- Cycle-time
- Value-added cycle-time (VA CT)
- Non-value-added cycle-time (NVA CT)

- Operators

Inventory

- Supermarket
- Buffer stock
- Number of Kanbans
- Wait time

The variables are factors influencing the essential value-stream-mapping timeline summary. Total cycle-time, total value-added cycle-time, total non-value-added cycle time, total lead time, total wait time, and process-cycle efficiency are commonly used process-summary metrics.[78]

KAIZEN EVENTS

Once the current-state value-stream map is documented, we start to identify areas of waste. Identifying Kaizen events, or "change for good," is an exercise that originated in lean manufacturing. This process is typically carried out over a week.

Five founding elements ground the event:

1. Teamwork
2. Personal discipline
3. Improved morale
4. Quality circles
5. Suggestions for improvements

All steps are important. However, for our purposes, we'll focus on the areas of improvement. Icons to represent the Kaizen events are added to the current-state value-stream map. The cross-functional collaboration event is used to gather data, analyze data, and identify waste and associated opportunities. Process maps, waste lists, or work layouts can be used.

The Gemba Walk, which refers to going where the work is, represents a common approach to walking the process. This way, each participant understands the end-to-end process by having walked through each step physically.[79]

The latter part of the Kaizen event is focused on running, measuring, and validating changes recommended. The changes affecting people are explained in detail and documented. Depending on the process being improved, a 5S method (sort, set in order, shine, standardize, sustain) can be useful, especially if the process deals with a workspace.

Once the recommendations have been presented, you're now ready to document the future-state value stream.

FUTURE-STATE VALUE STREAM

It's within or after the Kaizen events (depending on the company) that the team transitions from the current state to the future state. The critical outcome is to have a clearly documented picture of the desired future state.

Like our current state model, we must document the process steps and the details around each process step; e.g., cycle-time, wait time, etc.

Value-Management and Value-Stream Comparison Metrics

At this point, we have everything we need in place to start to get a better understanding of the effect the team had on improving the process.

We have three components to generate metrics for the value stream. First, the current-state value stream needs to be analyzed. Second, the future-state value stream needs analysis. Third, we can determine the difference between where we started in the current state and where we finished in the future state.[80]

The objective for the entire project, program, or initiative is to add value. A closely related objective is to quantify that value. This is the step where we do that.[81]

This step is primary in all lean initiatives, and we can use our value-stream map to address cycle-times and identify waste. It's useful to name the process steps intelligently, as those names will be reported out as-is in this step. Of course, they can be edited; however, it usually only introduces confusion when you're walking a new team through the quantified improvements, and between steps the names of activities are changed.[82] [83]

There are a few key variables to understand as you develop and calculate these values:

1. Seq ID
2. Shape
3. VA
4. NVA
5. Total

The Seq ID represents the number of shapes in your value-stream map. If there are only a handful of process steps, adding this might be unnecessary. However, once the model starts to have 10 or more steps and shapes, numbering references the shapes more precisely. The shape is the entity, by name, that's being referenced. The VA stands for value-added activities. The NVA stands for non-value-added activities. The total includes a sum of both the VA and NVA values.

CURRENT VALUE-STREAM METRICS

The current value-stream metrics are performed based on the current value-stream map that was created earlier. This example shows a simple contract process and where value-added activities and non-value-added activities were added. These metrics are focused on time.[84]

Seq ID	Shape	VA	NVA	Total
2	Draft SOW	14 hr	2 hr	960.0 min
3	Validate SOW	4 min	2 hr	124.0 min
4	Socialize SOW	2 hr	15 hr	1,020 min
22	Report SOW	5 hr	30 hr	2,100 min
21	Review SOW	8 min	30 min	38.00 min
	Totals:	21.20 hr	49.50 hr	70.70 hr
	% of Total:	30%	70%	

TABLE 1.2 CURRENT VALUE-STREAM METRICS

FUTURE VALUE-STREAM METRICS

The value-stream metrics are derived from calculations between the current-value-stream map and the future-value-stream map. These values represent the future state or, ideally, the new current state at this point (assuming changes are adopted). The example below shows similar metrics focused on the future-state value-stream map.

Seq ID	Shape	VA	NVA	Total
2	Draft SOW	6 hr	2 hr	480.C min
3	Validate SOW	4 min	1 hr	64.0C min
4	Socialize SOW	2 hr	1 hr	180.C min
21	Review SOW	2 min	1 min	3.00 min
22	Report SOW	2 hr	1 hr	180.C min
	Totals:	10.10 hr	5.02 hr	15.12 hr
	% of Total:	67%	33%	

TABLE 1.3 FUTURE VALUE-STREAM METRICS

VALUE-STREAM COMPARISON

Now that we have values calculated for both the current- and future-state value-stream maps, we can identify the difference between those two states.

The value-stream-comparison metric is an important metric to quantify the team's efforts and the results of the action taken.

This summary is often included with the final report to clarify that the project was delivered and, more specifically, articulates the value realized as a result of the initiative.

Performing this step helps to "make real" the efforts of the team. Looking at this another way, if the team didn't get involved, this is the quantified value that would have never been realized. This is a powerful statement and much more effective than a general narrative about abstract benefits achieved. Here's an example of what that summary might look like:[85]

	VA	NVA	Total
Improvement:	11.10 hr	44.48 hr	55.58 hr
% Improvement:	52%	90%	79%

TABLE 1.4 DELTA (CURRENT STATE - FUTURE STATE)

The percentage of work activity is another metric that's commonly leveraged. However, to keep the terminology more universal, we've used the term "percent improvement."

Lean Comparison Metrics

Lean comparison metrics are similar to value-stream comparison metrics. Yet, these metrics compare the baseline process map with the final process map.

LEAN BASELINE PROCESS MAP

The lean baseline process map reflects the state of our process prior to any changes being made. It's not always possible to obtain or develop this if the project or initiative is already in flight.

In either case, it can still be reasonably extrapolated and developed. It's ideal if this baseline process map is created before the project's transformation part kicks off.[86]

There are key fields worth understanding that are used in this process.

1. Seq ID
2. Step/Activity
3. VA
4. BVA
5. NBA
6. Total

Seq ID references the step number in the process. The step or activity refers to the name of the step or activity. VA is value-added activities. BVA is non-value-added activities that are required because of the business, systems, process, or technologies. NVA is non-value activities or waste that, ideally, can be eliminated. The total is the sum of these values. Here's an example of a lean baseline process map.

Seq ID	Step/Activity	VA	BVA	NVA	Total
1	Baseline Process as-is - Draft SOW	12.5 hr ▾	6.25 hr ▾	12.5 hr ▾	1,875.C min ▾
2	Baseline Process as-is - Validate SOW	480 min ▾	240 min ▾	480 min ▾	1,200.C min ▾
3	Baseline Process as-is - Socialize SOW	225 min ▾	450 min ▾	675 min ▾	1,350.C min ▾
4	Baseline Process as-is - Report SOW	270 min ▾	270 min ▾	270 min ▾	810.00 min ▾
5	Baseline Process as-is - Review SOW1	300 min ▾	150 min ▾	300 min ▾	750.00 min ▾
	Totals:	33.75 hr ▾	24.75 hr ▾	41.25 hr ▾	99.75 hr ▾
	% of Total:	34%	25%	41%	

TABLE 1.5 LEAN BASELINE PROCESS MAP

LEAN FINAL PROCESS MAP

The lean final process map is exactly like the baseline process map with the exception that it represents the final or future state (if changes haven't been implemented).[87]

The summary table below has the same fields as the baseline process-map table.

Seq ID	Step/Activity	VA	BVA	NVA	Total
8	Final Improved Process - Draft SOW1	2.6 hr	1 hr	1 hr	276.00 min
9	Final Improved Process - Validate SOW	80.4 min	30 min	30 min	140.40 min
10	Final Improved Process - Socialize SOW	120 min	120 min	120 min	360.00 min
12	Final Improved Process - Review SOW1	180 min	60 min	60 min	300.00 min
11	Final Improved Process - Report SOW	90 min	90 min	63 min	243.00 min
	Totals:	10.44 hr	6.00 hr	5.55 hr	21.99 hr
	% of Total:	47%	28%	25%	

TABLE 1.6 LEAN FINAL PROCESS MAP

The total and percentage of total reflect the value-added, non-value-added-but-required, and non-value-added totals for each process step.

LEAN PROCESS MAP DELTA

The lean process map delta takes the baseline process map metrics minus the final process map to determine variance. These calculations are the basis for quantified value of the initiative. Here's an example of the lean process map delta.

	VA	BVA	NVA	Total
Improvement:	23.31 hr	18.75 hr	35.70 hr	77.76 hr
% Improvement:	69%	76%	87%	78%

TABLE 1.7 DELTA (BASELINE – FINAL)

At this point, we've performed all the process analysis and calculations required to quantify the impact of the project in value-based terms for the value-management office.

Mastering Soft and Hard Savings

Determining the financial savings is the last step. Technically, financial savings are hard savings, and costs avoided are soft savings. However, for our example, we'll take a more generalized approach and classify savings simply as hard savings or soft savings.

A monthly savings estimation is done at the beginning of the project, and this is validated at the end. Often, this analysis is done at the monthly level. Here's an example of monthly estimations and actuals.

			Hard Savings		Soft Savings		Combined Savings	
#	Date		Estimated	Actual	Estimated	Actual	Estimated	Actual
1	Oct	2020	0	0	15,000	0	15,000	0
2	Nov	2020	25,000	12,500	15,000	0	40,000	12,500
3	Dec	2020	15,000	12,500	15,000	16,000	30,000	28,500
4	Jan	2021	15,000	25,000	15,000	16,000	30,000	41,000
5	Feb	2021	15,000	50,000	15,000	17,000	30,000	67,000
6	Mar	2021	15,000	75,000	15,000	13,000	30,000	88,000
7	Apr	2021	15,000	75,000	15,000	15,000	30,000	90,000
8	May	2021	15,000	75,000	30,000	15,000	45,000	90,000
9	Jun	2021	15,000	75,000	30,000	22,000	45,000	97,000
10	Jul	2021	15,000	75,000	12,000	25,000	27,000	100,000
11	Aug	2021	15,000	75,000	12,000	15,000	27,000	90,000
12	Sep	2021	15,000	75,000	12,000	15,000	27,000	90,000
	Project Totals:		175,000	625,000	201,000	169,000	376,000	794,000

TABLE 1.8 MONTHLY FINANCIAL SAVINGS

Once the monthly financial savings are calculated, we can then roll those numbers up and calculate the summary numbers. Although we quantified the process improvement as a result of the project's efforts, we still need to determine the economic benefit the project had on the organization. Here's an example of the financial value resulting from the project.

Annualized Hard Savings		Annualized Soft Savings	
Estimated	Actual	Estimated	Actual
175,000	625,000	201,000	169,000

TABLE 1.9 ANNUALIZED FINANCIAL SAVINGS

When presenting financial numbers, it's important to ensure that the information presented is accurate. Take the extra time to consult with your financial and corporate partners to validate the team's financial hard and soft savings.

Research Propositions

I developed the propositions that follow based on my research and the enormous potential to maximize the strategic value from the value-management office's efforts.[88]

Three propositions (P1, P2, and P3) are offered that have the potential to introduce impactful approaches to these concepts into an organization.

PROPOSITION 1: OPERATIONAL EXCELLENCE TEAMS

Depending on the organization's size, introducing such a radical framework to an operational and tactical project management office might be challenging.

Instead of taking a big-bang approach, it might be more prudent to identify an operational-excellence team that's trained to roll out strategic and organizationally complex initiatives.

This ensures that value is quantified and allows ideas and concepts to grow organically throughout an organization as they evolve toward a value-based framework.[89]

PROPOSITION 2: INTEGRATION WITH THE PROJECT MANAGEMENT LIFECYCLE (PMLC)

Not all organizations have the luxury of time. Critical deadlines, business commitments, and key milestones may prohibit a slower and more calculated rollout of this value-based framework.

In that event, integrating these key concepts into an existing project-management lifecycle will result in immediate benefits. While some degree of training and education would be required, that could happen through less formal communications channels such as webinars or computer-based training.

PROPOSITION 3: EMBED CHANGE AGENTS IN CRITICAL PROGRAMS

Another option for introducing these concepts into an organization more readily is to leverage change agents.

The idea of utilizing change agents is that a small group of individuals would be trained on these tools and techniques. These change agents would then be assigned and dedicated to a strategic program. The benefit of this approach is that educational overhead is decreased while still having representation within each of the strategic initiatives to ensure that value can be quantified.[90]

Limitations and Direction of Future Work

This chapter expands on the wealth of information already published on the subject of portfolio, program, and project management. The objective of this chapter is to present an alternative approach to simply executing projects. This alternative approach emphasizes value or volume.

Additional research is required to explain how other practical uses of value-based frameworks can be integrated into existing project-management offices that haven't been a benefit to companies.

The cultural shift this framework requires is significant and demands support from the executive leader to be successfully introduced and implemented.

Elevating Your Strategy with a Value-Centric Approach

Portfolio, program, and project leaders need to shift to value-management offices that are hyper-focused on strategic initiatives and the quantified value they produce.

Today, these leaders aren't trained, educated, or prepared to lead in a value-based world. Everyone is clear what value is—right up until they need to explain it for their project.

It's understandable but not acceptable to focus on overvaluing volume. It's not reasonable that value be measured on the number of production deployments when their benefit hasn't been quantified. Is releasing monthly better than quarterly, or weekly better than monthly? We don't know. An assessment must be done to quantify the value being delivered, but this rarely happens. We just assume that daily releases are better than quarterly releases, when that may not be the case.

Before we can expect changes from our teams, departments, and organizations, we need practical approaches to assist them in their transformation toward value-based thinking.

Change begins and ends with people. It's practical for executive leaders to provide a framework with boundaries and guideposts on reimagining their project delivery in a value-based world.[91]

Innovative companies are searching for new ways of operating. They want to be more agile. They need to deliver faster. They can't afford additional re-work. These organizations are looking for the best methodology to deliver value not projects. Changing our leadership, exploring new technologies, and shifting how work is performed all have the potential to transform organizations across the globe.

There's one common element regardless of the people involved, the systems used, or the technologies in play—people drive value. Employees need to be educated in value management to understand the difference between earned value management and value management. Value management can be applied to the simplest or most complex communication structures with dozens of teams, departments, and multiple companies working in collaboration.

Leaders that embrace value management as an essential organizational capability will transform their organizations and maximize organizational value.

CHAPTER 4

ASSESSING PROJECT MANAGEMENT COMPETENCIES

"My main job was developing talent. I was a gardener providing water and other nourishment to our top 750 people. Of course, I had to pull out some weds too."

- JACK WELCH

Chapter Objectives

After reading this chapter, readers will be able to:

- Understand how talent is linked to value
- Explore resources to define your standard of "good" performance
- Learn how to apply a framework to measure talent uniformly
- Determine how to benchmark your team to quantify improvements
- Learn how to link leadership decisions to quantified team improvements in competency

This chapter aims to present a practitioner approach to assessing project- and program-management capabilities across individuals, teams, and organizations. Executives must deliver value to business partners to stay relevant, engaged, and employed. Often, the value created doesn't equate to the net organizational investment. This situation creates challenges for the entire organization. Value isn't generated at a rate that sustains demand. When delivery is impacted, the expected project and initiative results aren't achieved, and value isn't realized. Many leaders blame individuals, focusing on the loudest but not necessarily the most critical areas. By applying a systematic approach to assess the management of an individual project or program, capabilities, the deficiencies within the team become self-evident.

As a result, fixing the problem becomes quite straightforward. This chapter will explain the method, approach, and technique to analyze and assess a team for demonstrated competency. Executive leaders will have a blueprint of what's required to effectively assess individuals, teams, and organizational capabilities to align to future business demands.

How Do You Define Success

When you're asked to build a team from scratch, it's challenging but refreshing for you as a leader. You set the *standard of good,* and you do the hiring. However, many leaders aren't that lucky. Most leaders are asked to take over a struggling team. This could mean assuming new responsibilities or simply replacing an existing leader that's not performing. Fixing a project-management team is difficult if there are a few gaps, but what if the gaps are everywhere?

The most common challenge for any leader, regardless of industry, is managing team performance. This could be a team of executives or a team of directors. The team might even be a group of developers. Each of those teams is managed differently. In this chapter, we're going to focus on project- and program-management teams.

Frequently, it's not only the team members' poor performance that's the problem. Leadership also needs to share responsibility for poor performance.

- Where do team members look for what's considered good?
- How are acceptable standards communicated?
- What are the areas of focus?
- How do project managers know if they're meeting expectations?

In this chapter, we'll assume a new leader has just inherited an existing team, and that team is perceived to be low-performing. Management wants people removed immediately. Projects are slipping. Risks are escalating at a scary pace. Fundamental design decisions need to be revisited. Projects near production need considerable re-work. Where do we start?

We begin with a complete assessment of our project- and program-management capabilities. It's inviting to start addressing problems as they fly by you, but this isn't sustainable. More importantly, it doesn't account for what we can't see.

Many planes were being shot down during bombing runs over Germany during World War II. Data showed damage to the fuselage (body of the aircraft) and wings of returning planes. The best approach? Reinforce the armor on the plane's fuselage and wings. The problem was this solution only included data from aircraft that made it home. The most critical aircraft hits were to the cockpit, engine, and parts of the tail—and these aircrafts never made it home. This solution, therefore, wouldn't work. The problem definition only considered the surviving aircraft and didn't consider aircraft that never made it home.

Similarly, the constant bombardment of requests and project fire drills creates tunnel vision on the in-flight or shortly starting projects—it doesn't consider projects that failed and no longer have fire drills.

When conducting capability assessments, we need to have a holistic picture of our capabilities outlined. We mustn't be tempted to only focus on the problems immediately in front of us using a subset of data.

Business partners want results. Unfortunately, if you're spending all your time as a leader dealing with fire drills, it's not easy to find time to access your departmental capabilities—but that's exactly what you need to do.

First, a few clarifications. Capability is defined as the ability to do something. At the same time, competency is the active possession and use of skills. This matters, because your team may, in fact, have the ability to improve the capability; e.g., they know how to build a detailed Microsoft project plan. However, they haven't exercised their competency to prove they can do it; e.g., build a project plan. It's much different if a team doesn't know they how to build a project plan as opposed to if they weren't clear a plan was required. We'll get deep into this discussion shortly.

We'll be taking a multi-step approach to quantify your project and program capabilities throughout this chapter. By following these steps, you'll have a systematic and repeatable method for assessing your team's demonstrated competency. We'll focus on project and program competencies; however, this methodology would work to evaluate a variety of roles, from business analysis to architects.

1. Identify capabilities
2. Define the standard of good
3. Assess demonstrated competencies
4. Benchmark
5. Provide feedback and report variances
6. Apply an organizational project-management-maturity model

By applying this framework, which I developed over the last 19 years, you'll have a clear understanding of where you need to focus your energy to course correct the team.[92 93]

Identify Capabilities Early to Maximize Value

We begin by identifying the core capabilities that we want to measure. It's essential to take time in this step. The decisions you make here will directly impact how your action plans are presented to your team and your executive leaders.

It's helpful to start by asking yourself a series of questions. What do I want to measure? Where has the team experienced challenges? What feedback do I consistently hear? If I could fix one area, where would I start? After running through these types of questions in your head, start to elaborate on the categories you'll use to frame the assessment questions.

There are no correct or incorrect answers here. However, what you choose in this step will significantly impact the scope of your action plan.

I'll offer a place to begin. One logical starting point is to look at project managers and program managers as separate roles. These roles might indeed be near similar in your organization. Yet, as a discipline, they do have distinct differences. This is a great starting point to understand better what work is getting done.

Start with identifying 10 standard areas of focus for *project* managers. These areas are hot spots within which all work, activities, and tasks fall:

1. Integration management
2. Scope management
3. Time management
4. Cost management
5. Quality management
6. Human-resource management
7. Risk management

8. Procurement management
9. Value and stakeholder management.
10. Communication management

As you may recall, these are standard Project Management Institute knowledge areas. While these do ebb and flow each year with the new PMBOK revisions, largely, they remain consistent year over year.[94]

Likewise, next, identify 10 standard areas of focus for *program* managers. These standard areas will be slightly different from the project management areas we defined earlier. I developed the program-management areas below based on my 19 years leading project, program, and portfolio delivery and transformation teams:

1. Program-portfolio management
2. Strategy and investment funding
3. Program governance
4. Program-contract management
5. Portfolio metrics and quality management
6. Human-capital management
7. Program-value and risk management
8. Program-management financial performance
9. Program-stakeholder management
10. Communication management

At this point, we've clearly identified our capabilities for both project and program managers. I won't be explaining each, as it's logical to extrapolate the intent of each ability.[95]

The leader must establish a quantified current state before any changes are introduced or implemented. Without a current-state assessment, the value or impact of future changes can't be quantified. Executives want quantified results.

Define the Standard of Good

The benefit of developing a standard of good is to demarcate a clear line between good performance and poor performance. Without communicating this definition to the team, there will be the ambiguity of what's expected. This lack of clarity will ultimately manifest as excuses for poor performance in the future.

The standard of good is the collective knowledge of best practices or principles that you, as a leader, have established to reinforce positive outcomes.

There are many great examples of good standards, and I'll highlight some of the more effective ones:

1. PPM rubric: explicitly describes the levels of performance for a role
2. Project financials: the source of truth for financial management on a project
3. Resource management and capacity planning: records how resources are being managed on a given initiative
4. Milestone peer reviews: colleague-based reviews of work products prior to submission
5. Project manager checklist: a master list of activities that would highlight best practices and the timing of activities for a project manager

Each of these standards of good reinforces that guidelines exist to determine good versus poor performance. They also help team members answer questions: Do I need to perform this activity? When do I need to perform this activity? These are some of the questions a clear standard of good helps to clarify.[96]

Team members will use these documents as references for best practices. The more accessible and straightforward you can make these documents, the easier it will be for your team to adapt and comply with best-practice standards.[97]

Assess Demonstrated Competencies

The capabilities are defined, and the standards for good are clear to the team. We're now ready to assess the demonstrated competencies.

Ideally, the standards for good would be defined first. However, if you find yourself stepping into a new team, department, or organization, and those standards are unclear or, frankly, don't align with your standard for good, that's okay. Use your existing experience to assess the team and work collaboratively with the team to get those standards for guiding principles.

There's a significant difference between asking a project manager if there's a project plan for their project versus asking if they know how to build a project plan. This is where most assessments fall short.

If I asked you how fast you should drive on a road with a speed limit of 65 mph, you'd be quick to parrot back an answer of 65 mph. However, if we reviewed your speed the last time you were on that same road, we might find you drove more like 75 mph. This is similar to why a demonstrated competency assessment is effective.

We're assessing whether the resource has demonstrated the skill (in the present), not whether the resource understands how to perform the skill. As a result, this style of assessment is curt and succinct. There's no latitude for excuses.

The demonstrated competency is a rapid-fire set of questions, with the resource sharing their screen to show they performed the skill. To help prepare, here's a quick guideline of tips and recommendations when conducting an assessment:

- From the resource's response, interpret, and update the status with these acceptable answers: Yes, No, or N/A.
- Fill out actions required after the assessment.
- Have the resource share their screen to validate the application of the skill.

- If you don't see what you need after 15 seconds, mark this question as "No," and move to the next question.
- Limit people in the room during the assessment. This helps minimize distractions.
- Only one person (typically a supervisor) should be asking the questions for the full assessment.
- Allocate 90 minutes for each assessment.

This quickly brings us to which questions you should ask. I'll provide a few examples. Generally, asking seven to 10 questions per capability area is adequate. The person leading the assessment will need to generate questions by capability. This helps to ensure that the resource understands the area of focus. Here are some sample questions for a project-manager assessment:

- Integration management: What projects have upstream or downstream dependencies with this project? Where are they captured?
- Scope management: Explain where I'd find the business case for this project. Show me.
- Time management: What does the team use to track progress? Show me.
- Cost management: Where do you track monthly burn for this project? Show me YTD numbers.
- Quality management: How do you personally ensure the quality assurance of this project? Show me what's being used to track quality effectiveness?
- Human-resource management: If you couldn't go to work tomorrow, where would the business analyst find the team vacation schedule? Show me where that's located.
- Risk management: What are your active risks vs. issues on this project? Where's the mitigation documented?

- Procurement management: Where's the statement of work stored for this project? Show me the latest version of the approved statement of work.
- Value and stakeholder management: What's been committed to your business partners? If you leave for a bigger project tomorrow, what am I looking at to validate your assumptions?
- Communication management: Your business analyst is a no-show. Who's the backup? Where do you find this out? Show me.

These are sample questions that address the intent to validate that resources are thinking through potential risky situations. Similar but different questions are asked of your program managers. There's overlap with a slightly broader scope in these questions. Here are some sample questions for a program-manager assessment:[98]

- Program-portfolio management: How is the authority to start work granted? How is this documented?
- Strategy and investment funding: What program opportunities are expected to result from this project?
- Program governance: How do you confirm resources are allocated correctly to your sprints, releases, iteration, etc.?
- Program-contract management: How many contracts are involved in this current project? Show me, in the contract, where deliverables are tied to payments.
- Portfolio metrics and quality management: Show me the peer reviews conducted to date on this project.
- Human-capital management: Where are project roles and areas of accountability described and documented?
- Program value and risk management: Is this project linked to a strategic goal? Show me where this is captured and documented.

- Program-management financial performance: What % are you burning against your project as of today? Prove it with data.
- Program-stakeholder management: Where are critical success criteria defined and agreed upon with the stakeholders for this project? Show me.
- Communication management: What information is communicated to project leaders on a weekly, monthly, and quarterly basis? Show me your last report.

Clearly, this is only a light sampling of questions that would be required to conduct a successfully demonstrated competency assessment. In reality, another seven to 10 questions per capability would be prepared.

As you progress through the questions, all you record is your interpretation of the status (Yes, No, or N/A), a brief comment, and action required (if time allows). This makes for a very fast and efficient assessment. This data is then rolled up to establish the score for the capability. To keep the assessment model simple, we're not going to elaborate on the process to calculate responses within this chapter.

It also doesn't matter where the team scores today. All you need out of the assessment is a benchmark of where they are today in terms of meeting your standard of good. The approach moves us from a narrow focus to a broader focus that indicates which capabilities have deficiencies and which capabilities have demonstrated strength.

This exercise results in the creation of radar charts for each of the project and program managers on the team. This way, they can visually see where there are deficiencies between their demonstrated competency and the expected organizational standards.

Benchmark Individual and Team Performance

However, before we share this information with the resources, we'll need to aggregate the team information to determine the team norms. Once we have the average performance levels, we'll show the resources where they are in relationship to the team's performance on each of the standards.

For example, in Chart 1.0, we charted the performance of project manager #1 during quarter one and quarter two. This visual contrast makes it very easy to determine which areas require improvement.

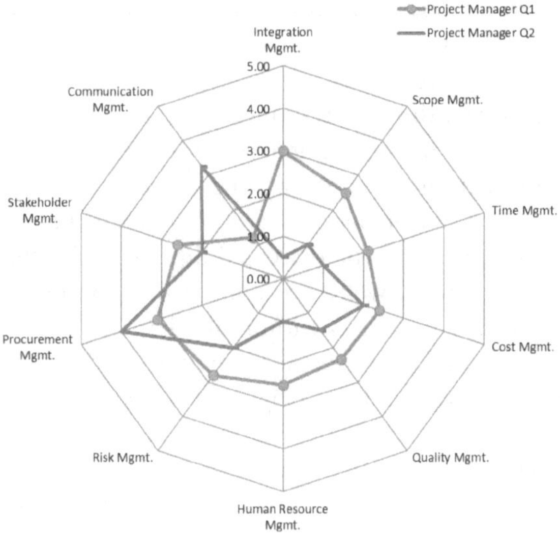

CHART 1.0 INDIVIDUAL Q1 VS. INDIVIDUAL Q2 PERFORMANCE

In Chart 1.1, we've summarized the results of all the project managers and determined the average performance, indicated in light blue. We've also charted the performance, from the demonstrated competency assessment, of project manager #1. As you can see, the performance of project manager #1 is considerably below the team performance in all areas except procurement management.

This chart will be part of the feedback that we provide to the project manager for their action plan.

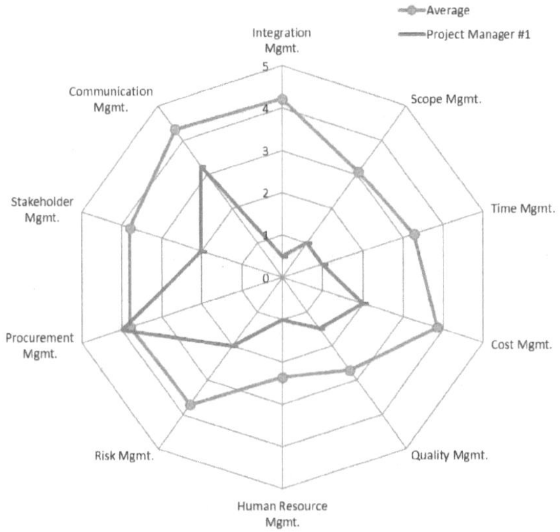

CHART 1.1 INDIVIDUAL VS. TEAM PERFORMANCE

By making the results visible, the delta in performance between the individual and the team is more apparent.

We've conducted the assessments and have the collective data. Our next step is to start to make a benchmark. Creating a benchmark indicates how mature the team is as of the date of the assessment. In short, the benchmark establishes the first pillar required to determine how well the team measures up against your standard of good.

You can conduct these assessments at whatever frequency you feel necessary. However, I've found that performing them quarterly is a great frequency to ensure team members are progressing and haven't let essential skills atrophy.

After the initial benchmark is established, you start to focus on areas of weakness. Using your standards of good and the action plans, you begin to create a picture of where effort needs to be focused for a quantifiable change to occur.

Once you have the benchmark results and implement new changes or processes or offer new guidelines, conduct another assessment. At this point, we have data from our initial assessment and our quarterly performance results. We're now ready to quantify our data and glean some insights.

While the individual-vs.-team demonstrated competency highlights areas related to the individual, the quarter-over-quarter demonstrated competency highlights changes in the team's performance over time. This is especially useful when articulating the value achieved and performance improvements of a team over time. This is an excellent tool to quantify results your leadership achieved after setting the standard of good, benchmarking, assessing, developing action plans, and measuring the impact of those plans on the team.

Chart 1.2 illustrates the collective feedback from all program managers for Q2 2020 compared to Q3 2020. This is a great tool for demonstrating the effectiveness of your action plans to improve performance.

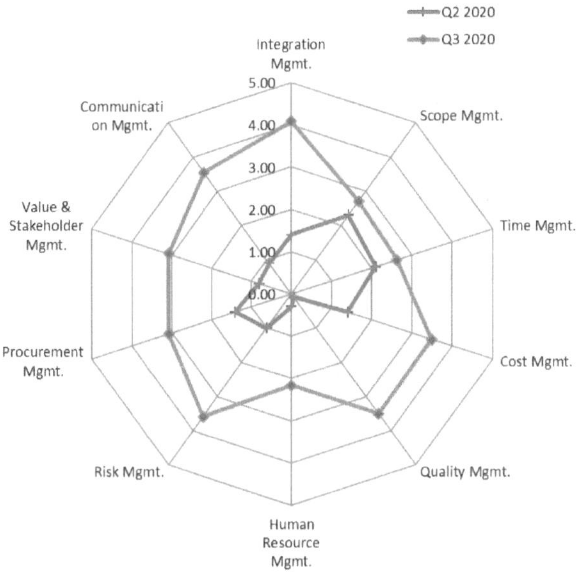

CHART 1.2 TEAM QUARTER-OVER-QUARTER PERFORMANCE

Provide Feedback and Report Variances

Benchmark results are complete. We have the required data to process analytics on individuals. This is one of the more impactful steps of this process. It clearly identifies where an individual is doing well and where they need to improve.

Instead of providing this information separate from the team's performance, we provide this feedback complemented by team performance. Most individuals don't want to drag down the team. As a result, this process is very useful for stimulating motivation to change the team's values, beliefs, and behaviors, resulting in improved performance. There are three pieces of information provided to each individual project manager:

1. Individual vs. individual last-quarter performance
2. Individual vs. team performance
3. Team quarter vs. team last-quarter performance

This format works extremely well to encourage change. By using this format, it's clear what we're using for the standard of good and how the project manager compares to that standard. Additionally, it's directive in highlighting where the project manager needs to focus to ensure they're within the team norm during the next quarterly demonstrated competency

Typically, the last step is to provide a ranked view of the team. Often, it's helpful to have a ranked list with some content. Specifically, categorize the team into three buckets:

1. Invest
2. Stable
3. Review

"Invest" is tagged to the top 5%-10% of the team. These are individuals that truly excel. "Stable" are individuals that are meeting the standard of good for the team. These are what we'd consider solid B or B+ players. Lastly, we have the "Review" tag. This is used for project managers or individuals that are substantially below the standard of good performance. These individuals are at risk. They'll either be put on a performance-review plan or, if contractors, rolled off the project rather quickly.[99]

It's also important to recognize the top performers. This helps the team understand who they can go to for help in areas where they're weak. This sets up the environment for a collaborative and healthy team dynamic.

Here's a full example of what the feedback format for an individual looks like:

To: Jane Smith,

During the period of Q1 2020, you performed 5% below the team average of 75% and fell within the 62nd percentile among the team.
Overall, the team quarter-over-quarter has improved by 8%, to 75%.
Specifically, the below analysis outlines how you when compared to the team average for this quarter.
Individual vs. team performance for the quarter:

Integration Mgmt.:14% vs. 94%, below target
Scope Mgmt.: 20% vs. 60% below target
Time Mgmt.: 22% vs.83%, below target
Cost Mgmt.:24% vs. 78%, below target
Quality Mgmt.: 35% vs. 75%, below target
Human Resource Mgmt.:18% vs. 74%, below target
Risk Mgmt.: 42% vs. 88%, below target
Procurement Mgmt.: 87% vs.90%, near target
Value & Stakeholder Mgmt.: 43% vs. 88%, below target
Communication Mgmt.: 64% vs. 94%, below target

<Insert Radar Chart: Individual vs. individual last quarter performance>
<Insert Radar Chart: Individual vs. Team Performance>
<Insert Radar Chart: team quarter vs. team last quarter performance

Thanks,
Leadership

**Note: Data is representative and not fully tied to calculations, as would be the case in a fully demonstrated competency assessment.*

Apply an Organizational Project-Management Maturity Model

The prior steps will provide a detailed and actionable plan to transform your group into a best-in-class team of high-performing project and program managers.

What's the next step? Does this continue quarter after quarter forever? How do we continually improve? I'll explain some of my best ideas for designing innovation into your demonstrated-competency assessments in research propositions.

There's been a lot of great research over the last 20 years focused on developing organizational maturity. Ken Crawford was the President of the Project Management Institute (PMI) from 1994 to 1996. Shortly after this time, in 2001, he authored the book, *Project Management Maturity Model, Edition 1*. This book is now in its third edition, published in 2014. I was fortunate to be involved in the Project Management Institute around the same time and was Project Management Professional (PMP) certified in 2001. This book serves as a great reference for evaluating project-management maturity.

Many folks who weren't involved with project management 20 years ago incorrectly assume these models were generated from consulting firms. In reality, Crawford envisioned and developed the foundation of the modern project-management maturity model.

Crawford expanded upon the foundational five levels of project management maturity:

1. Level 1: Initial process
2. Level 2: Structure process and standards
3. Level 3: Organizational standards and institutionalized process
4. Level 4: Managed process
5. Level 5: Optimizing process

Crawford also defines the eight components of the project-portfolio maturity (PPM) model.[100] [101]

1. Portfolio governance
2. Project opportunity assessment and initiation
3. Project prioritization and selection
4. Portfolio and project communications management
5. Portfolio resource management
6. Portfolio risk management
7. Portfolio-management organizational structure
8. Portfolio performance management

Each maturity in Levels 1 through 5 is assessed for each of the eight components. The general process to evaluate PPM capabilities involves the following steps:

1. Analyze departmental-portfolio maturity level.
2. Analyze portfolio maturity at the enterprise level.
3. Determine the enterprise-portfolio maturity level.
4. Establish target-enterprise and departmental-portfolio maturity levels.
5. Expand the desired state.
6. Define current state.
7. Create an action plan.

Performing these steps will help ensure that your portfolio maturity model is fit-for-purpose and is aligned with your enterprise strategic vision.[102]

Analyze Departmental-Portfolio Maturity Level

Begin by having executives conduct a portfolio-maturity analysis on a single department, division, or region. Be sure that the feedback and data are from business partners, sponsors, executives, and portfolio and program managers who work in the environment. This will ensure that the data is not skewed or fraught with confirmation bias.

Table 1.0 is an example of what an initial assessment might look like. This is a quick-and-dirty perspective on maturity across the department.

It's important to note that specific measures, activities, and processes must be set up and established at each level to allow the determination of red, yellow, or green. For simplicity, we won't cover the detailed explanations of those interpretations. An experienced portfolio executive will have frameworks that outline these levels and standardization criteria in detail.

However, I'll provide example criteria from Level 1 portfolio-resource management as a frame of reference:

- A single repository for aggregating departmental resource and demand are utilized.
- Demand vs. capacity management can be analyzed on demand.
- Department has a defined process for resource management.
- Resource management is operationalized across the department.
- Resource demand can be analyzed for a one-year look-forward window.
- Roles and responsibilities for portfolio governance at each organizational level aren't defined, or the roles and responsibilities are inconsistent throughout the organization.
- Documentation detailing roles and responsibilities doesn't exist.
- Roles and responsibilities aren't aligned with identifying and managing project opportunities.
- Resources are assigned to projects via ad hoc processes.
- There are no inventories of resource skills or assets to use in assignment processes.
- Documentation of resource-management processes doesn't exist

- Resources are assigned to projects on an availability basis using ad hoc processes.
- Documentation of the processes may or may not exist.
- Delays in securing resources for staff projects are common.
- Reliance on limited, highly skilled labor resources that are over-allocated across multiple concurrent projects is common.
- No skills inventory or asset inventory exists to depict the organization's resource capabilities.
- Resource skills and availability data aren't documented or readily available for assessing resource requirements for potential, pending, or active projects.
- Nonexistent or rudimentary processes exist to plan resource requirements to meet near-term or future portfolio needs.[103,104]

The assessment can be completed rather quickly by evaluating these criteria as well as that allocated to the other component areas. Table 1.0 is representative of what a completed, summarized, departmental, portfolio maturity-model assessment may look like.

	Level 1	Level 2	Level 3	Level 4	Level 5
Portfolio governance	G	A	A	A	R
Project opp. ass. and initiation	A	R	R	A	R
Project prior and selection	A	A	A	A	R
Portfolio & project comm mgmt.	G	A	R	A	R
Portfolio resource mgmt.	G	A	R	R	R
Portfolio risk mgmt	G	A	R	R	R
Portfolio mgmt. org structure	G	A	R	R	R
Portfolio performance mgmt.	G	G	A	R	R

TABLE 1.0 DEPARTMENTAL PORTFOLIO MATURITY MODEL ASSESSMENT

Analyze Enterprise-Portfolio Maturity Level

Needless to say, the enterprise-portfolio maturity level will be a rolled-up view of the combined departmental-portfolio maturity levels.

Quite often, the rolled-up view is much less mature than many of the departmental-portfolio maturity assessments. Enterprise PPMs will look very similar to the departmental model.

For the enterprise-portfolio maturity model to represent the whole, it's important to validate that departmental models were conducted only based on their understanding of the departmental maturity and not that of the enterprise. Often, instructions are unclear, which results in inconsistent data that makes aggregating and pulling insights from the data difficult.

	Level 1	Level 2	Level 3	Level 4	Level 5
Portfolio governance	G	A	R	R	R
Project opp. ass. and initiation	G	R	R	R	R
Project prior and selection	G	A	R	R	R
Portfolio & project comm mgmt.	G	A	R	A	R
Portfolio resource mgmt.	A	R	R	R	R
Portfolio risk mgmt	G	A	R	R	R
Portfolio mgmt. org structure	A	R	R	R	R
Portfolio performance mgmt.	G	G	A	R	R

TABLE 1.1 ENTERPRISE-PORTFOLIO MATURITY-MODEL ASSESSMENT

As we can see in Table 1.1, the general-component-level maturity is less mature. This is to be expected. Additionally, some areas, when combined with other departments, increased in maturity (e.g., Level 1 of project prioritization and selection) while other components were at lower levels of maturity (e.g., Level 3 of portfolio governance).

Another useful view is showing each department separately, highlighting the departmental differences in maturity.

	Level 1		Level 2		Level 3		Level 4		Level 5	
	IT	HR	IT	HR	IT	HR	IT	HR	IT	HR
Portfolio governance	G	G	A	R	R	R	R	R	R	R
Project opp. ass. and initiation	G	A	R	R	R	R	R	R	R	R
Project prior and selection	A	A	G	A	R	A	R	R	R	R
Portfolio & project comm mgmt.	G	G	G	A	R	A	A	R	R	R
Portfolio resource mgmt.	A	R	R	R	R	R	R	R	R	R
Portfolio risk mgmt	G	G	A	A	R	R	R	R	R	R
Portfolio mgmt. org structure	A	G	R	R	R	R	R	R	R	R
Portfolio performance mgmt.	G	G	G	G	A	A	R	R	R	R

TABLE 1.2 ENTERPRISE PPM COMPARISON

This view can be useful to narrow the differences among departments quickly. For example, we can determine that IT is more mature than the HR department. We can also deduce that resource management, as a capability, is immature for both departments. We can then deduce that these departments likely don't have much confidence that their year-one or year-three strategic plans will be achieved, given that they can't prove resources are properly staffed to fulfill those commitments.

The models are straightforward yet still help highlight potential areas of focus that could be addressed.

Establish Target-Enterprise and Departmental-Maturity Levels

Observing these models, it's easy to assume the most mature portfolio level is Level 5. You'd be correct. However, it's also easy to assume that all enterprises and departments desire to achieve Level 5 maturity. This is simply not the case.

With maturity comes overhead, processes, standards, and procedures. Often, this is all wrapped up with procedural controls mandated and controlled by systems. If the department is in a highly regulated industry, this could be a good thing. However, if that enterprise focuses on innovation or research, maturity becomes a restriction that prevents rather than accelerates progress.

A quick analysis of our departmental-portfolio maturity assessment indicates that the department has achieved Level 1.7. Alternatively, when observing the enterprise-portfolio maturity-assessment data, the enterprise has reached Level 1.2.

Next, we need to determine our desired PPM maturity. Let's assume our company is a progressive and innovative healthcare company offering cutting-edge SaaS solutions for customers. We do want documented processes; however, we don't want to limit our ability to be flexible for our business partners. As a result, we evaluate all the options and determine that Level 3.5 portfolio maturity is appropriate for our department. For simplicity, we'll assume the Level 3.5 desired portfolio maturity target lines up with our desired enterprise portfolio-maturity level targets—though, in reality, this may not be the case.

We do have a gap. However, it's not as drastic as if we had a Level-5 maturity target.

Expand the Desired State

The desired state represents the ideal future state of maturity. If the ideal state is Level 3.5, be sure not to pull criteria from a Level-5 portfolio that's fully optimized with defined processes.

The table below outlines what a simplified desired state for a Level 1 initial process may look like.

Project Prioritization and Selection	Just happens, project selection process is undocumented and inconsistent	G	• Formal process defined for project selection • Single source of truth for requested projects • Projects ranked (weighted) based on anticipated scientific value recognition • Initial financial swag provided +/- 50% • Stakeholders are defined

TABLE 1.3 LEVEL 1 INITIAL PROCESS FOR THE DESIRED STATE

The intent here is to identify the target state for each level of the five levels and against all eight components. If not already developed by the executive leading the initiative, this will be a time-consuming process.

Define the Current State

The current state explains each component that's already in place and activities that meet or exceed that level's standards and components. Essentially, you're explaining what's in place today. If there are limited pieces or there's nothing in place today, you document the gaps at this stage.

The table below documents the component details that are in place today for this level.

Portfolio Resource Management	Ad hoc resource management occurs at the project level	G	• Resource management rolled out to department in early 2018 • All FTE and consultant resources loaded and tagged to projects • Evolving, directionally correct data here • Standardize approach for risk and issue management • Risks and issues loaded into PPM tool weekly • Monthly meetings conducted to review resource variance to project plans for the department

TABLE 1.3 LEVEL 1 INITIAL PROCESS FOR THE CURRENT STATE

Suppose this step of documenting the current state hasn't been completed. In that case, there won't be a clear action plan to close the gap between the departmental-portfolio maturity level and the target-departmental maturity model. The same logic applies to the enterprise-portfolio maturity model. Often, there's a lot of reuse between the departmental- and the enterprise-maturity model current state.

After you have the desired state and the current state, you're now in a position to outline an action plan to close this gap.

Create an Action Plan

Given that we have the current-state plan in place, it's rather quick to extract the action plan from that existing analysis.

Start by removing all the green status components. By removing these components for each level that's 100% green, we establish our action plan.

The table below offers an example of what an informal action plan would look like. There's a report of business cases in this example, but the executive leaders aren't using the reports. We could argue that it's partially done—possibly why it's yellow—but, if it's not used, it's not adding value.

Portfolio Governance	Process exists to create portfolio review boards Board integrates project management & business knowledge –relates projects to strategies	A	• PMO has a generated report of business cases. This is useful. Project level views are not utilized by teams of leaders (as perceived as not useful) • We do have all projects linked to strategic initiatives (12 strategic projects) and are reporting to VP level this information quarterly (and often ad-hoc, or when new funding needs arise).

TABLE 1.3 INFORMAL ACTION PLAN

Applying a similar logic to the demonstrated-competency assessment, we can apply those techniques as we tighten up the portfolio maturity.

Using a measured approach provides a visible standard of what good performance looks like for the team. Additionally, by making the expectations clear, we focus on maturity in areas where we need it and prevent the process from impacting our organizational ability to innovate.

Research Propositions

The purpose of the following propositions isn't to pontificate about things that could be done but rather identify areas that could create additional value to the demonstrated competency assessment if expanded.[105]

I developed these propositions based on my research and assumed that leveraging a portfolio-maturity model can offer benefits for team alignment with strategic goals.

Often, there's weak guidance on how to conduct these maturity assessments. This creates differences among the parties that complete the results and, therefore, introduces unnecessary variability. Three propositions (P1, P2, and P3) are offered that have the potential to create a reference model for departmental or enterprise-capability improvements.

PROPOSITION 1: START, STOP, AND CONTINUE

Usually, an assessment is done because of a perception of low performance. Rarely are assessments requested when performance is exceptional.[106]

Often, leaders fail to define good before taking action. When absorbing a new team struggling to perform or recharging an existing team, don't ask everyone on the team what activities they perform. At best, you'll document the current state of activities that define low performance; at worst, you'll miss the real activities not being performed that directly contribute to the team's dysfunction. You must first define good.

Interviewing resources (project managers, program managers, and portfolio directors) to find out what they're working on doesn't add value. However, after you've defined what good looks like, you do need to compare what activities resources are performing and bump that against your definition of good.

First, in my article, *Define "good" to scale your organization*, on CIO.com, I explain my method to do just that—define good. However, you can add the start, stop, and continue technique.[107]

It's as simple as it sounds. What activities should be started that haven't been focused on? Which activities should be stopped immediately, as they don't add value or are out of the role's scope? And, lastly, which activities or behavior should be continued, because it's adding value.[108]

In most situations, you'll find that a significant part of the work your team is doing is actually being performed to compensate for other teams' deficiencies.[109]

PROPOSITION 2: ESTABLISH AN ENTERPRISE-RESOURCE-CAPACITY AND DEMAND-MANAGEMENT MODEL

It's not always possible to complete all requests in a single day, week, month, or year. The more senior you get in your role, the more status quo it will be to have loads more work than you could ever finish—or even more than the entire team could reasonably complete.

Without a view into how resources are allocated over time, it will always fall onto the leader's shoulders to "find a way" to make the staffing model work. That is, unless you prove it's impossible, and additional resources are required.

There are many approaches to effectively perform enterprise-resource planning, so we won't cover that in depth here. However, I will offer that after completing a resource plan for all active or soon-to-be-active projects—with all necessary resources from project managers to developers—this exercise almost always results in a reprioritization of the portfolio.

Validate that your team is positioned and staffed correctly for success, and start by developing an enterprise-resource-capacity and demand-management model.[110]

PROPOSITION 3: ACTIVELY MANAGE INTAKE AND REQUEST MANAGEMENT

You've probably been in a situation where your team is getting what seems like constant new requests. Existing requests aren't even completed yet. They could be executive decks, issues to resolve, or prioritizing strategic projects. The activities in question really don't matter. What does matter is that every request exceeds your team's capability to deliver it.[111]

This results in a flurry of additional requests for the status of the old requests that, of course, aren't yet completed. This turns into a vicious cycle and usually results in the removal of the manager from the situation. Let's get ahead of that worst-case scenario.

The solution is to establish an enterprise or departmental process for the intake of requests for your team.

A product like ServiceNow could be used, or it could be a simple, robotic, process-automation script that scrapes a generic inbox, or a tracking worksheet in Excel. The method doesn't matter as much as the act of queuing requests. Why is this critical? This process provides insight into the volume and scope of demand your team supports—or can't support, as the case may be.

You now have a complete picture of the full scope of intake requests and how those requests impact your team's utilization. Once you have this view, you have the basis for creating service-level agreements (SLAs), operational-level agreements (OLAs), and business-level agreements (BLAs).

At this point, you've defined your team's level of service. For example, it may take five business days to conduct a sprint-planning exercise with your currently staffed team. You can demonstrate that, today, you have a five-day SLA for sprint planning, but, with an additional resource, this could be compressed down to three days, a 40% efficiency improvement.[112]

With this approach, you're not suggesting adding another resource, but you're stating, matter of factly, that, with the current team, it's going to take, on average, five days to complete these types of requests.

Now your team is managing expectations it can reasonably meet. This is much more motivating for your team.

Limitations and Direction of Future Work

This chapter extends the voluminous research area dedicated to capability maturity models within the context of project, program, and portfolio management. This chapter aims to present ideas to address how to improve project, program, and portfolio performance by applying a repeatable, measured, and quantified method.

Additional research is required to elaborate on other techniques that could be leveraged to produce similar results in less time. As this technique scales, it requires a serious time commitment. The results stand on their own as substantive. However, the additional workload required to support a team of 50 vs. 500 is substantial.

It would be useful to see additional research published that addresses approaches other than a divide-and-conquer approach through delegation to produce similar quantified results.

Why Measuring Change Matters

Change can be measured. Whether you're on a racetrack, hiking a mountain, or even improving a team, change is measurable. It's great to have a vision for the future. Almost every leader has a vision, but not all leaders realize their vision. Sustainable change must be measured to track progress and prevent regression.

It's exciting to lead a new team. The potential is limitless, and it's fun to consider the possibilities, the relationships you'll build, the partnerships you'll enable, and the team members you'll empower. However, for any of this to be realized, you need to improve the maturity so risks aren't escalated daily and fire drills don't interrupt standing meetings.

By taking a thoughtful approach and applying a demonstrated competency model, you have the framework to measure and quantify the impact your team has made while under your leadership. This is very impactful, as you're now able to specifically articulate what change occurred and tie it to outcomes or value realized.

Think back to the teams you've led over the years or the teams you've been a part of. Were those teams able to quantify their results and impact on the organization? Likely, process improvements were made or new technologies were introduced, but those impacts (the value) weren't quantified.

This is the primary benefit of a demonstrated-competency model. We're evaluating whether the resource is applying the skill today, in the present, not whether they understand the theory behind it.

This shifts the discussion from a social popularity contest to a clear, black-and-white, action-based performance environment. Is the resource performing the expected behavior or not? If so, encourage more of the same and open new doors. However, if the resource isn't performing to the standard, ensure the resource has knowledge, tools, and techniques to perform at the level expected.

When everything is going wrong in a department or enterprise, it's difficult to see the root causes. Frequently, in all honestly, you might not find them initially. This is why taking a holistic approach to performance management and applying a demonstrated competency model is effective.

The next time your team is working 50-60 hours a week and still not getting the results expected, consider performing a project- and program-management demonstrated-competency model. The results may surprise you.

CHAPTER 5

LEVERAGING SERVICE CATALOGS TO COMMUNICATE VALUE

"The more you focus on the value of your product or service, the less important the price becomes."

- BRIAN TRACY

Chapter Objectives

After reading this chapter, readers will be able to:

- Observe your organizational portfolio, program and project offerings from a service catalog perspective
- Explain a best-practice service offering for your department
- Analyze your department's value proposition from a service catalog perspective
- Apply useful metrics to service categories
- Learn the benefit or reframing value from the perspective of your business partners

This chapter aims to present a unique approach to implementing a project-management-office service catalog for organizations focused on continuous value delivery. Business information executives and officers must validate investment decisions and demonstrate value achieved. The process of gaining additional organizational and cross-functional executive buy-in is especially difficult when team members, leaders, and executives don't understand what services the organization's project-management office provides. The traditional project-management office—centered around processes and templates—is being transformed into a new-age, value-management office that's hyper-focused on the value realized. Innovators understand and appreciate that if the services the project-management office provides are vague, business partners won't consume them.

The act of creating a service catalog allows for a tailored or agile approach to delivery. This model accepts that not all projects are created equal. More specifically, leveraging a value-management service catalog refocuses the organization. Core capabilities are provided through the catalog, and investments in these areas are doubled down. Alternatively, capabilities that fall outside the core capabilities of the value-management office are evaluated for outsourcing. This shifts the traditional project-management office from a cost center to a value center and makes services offered transparent to downstream consumers.

Why People Need to Understand Your Offerings

The traditional project-management office is encapsulated within the newly formed value-management office. This new structure focuses on value achieved and offers greater flexibility to business partners.

Typically, the project-management office was presented as a best-practice entity to advance maturity in the organization. While conceptually sound, in practice, this structure quickly breaks down. Efforts shift from why we're doing a project to how to do more projects. In this effort to deliver more with less, the "why" of executing the project moves into the background.

Business partners realize the project-management office isn't providing value. All too often, the project-management office is creating templates that no one cares about. They check boxes off spreadsheets that are maintained for no apparent reason. The project-management-office agents—the portfolio, program, and project managers—stop thinking. They become robots of compliance to policies that are vague, standards that don't reflect reality, guidelines that are misaligned to business partners and mandates for outdated procedures, and then project-management-office leaders are surprised when support for the PMO wanes. Each subsequent PMO hire becomes more and more difficult. We shouldn't be surprised.

Innovative leaders acknowledge a need to maintain updated policies, templates that reflect reality, and procedures that align to business goals. These maintenance and operational functions help the company deliver a project using best practices, but they don't prevent the project from being completed without adding any value. This is the function of the value-management office, which is hyper-focused on ensuring the most strategic company projects are delivered with excellence and maximum value.[113]

Before a value-management office can be launched, the company must first determine the office's core capabilities. Core capabilities will become the priority for staffing, and non-core capabilities will be supported but likely outsourced. The distinction between core and non-core is vital.

Here's the rationale. When presented with additional work, most executives will develop an algorithm—something to the effect of: 100 people produce 20 projects, and 200 people produce 40 projects; therefore, to do 100 projects, I require 500 people. The base premise is that more work requires more people. This can be true. It also can be false. Instead of asking for another 10 or 100 resources, the value-management office allocates resources more efficiently.

Often, the waste (a result of inefficient processes) is bucketed into three classifications: (1) value-added; (2) non-value-added, required; and (3) waste (or non-value-added, not required). Value-added activities produce a positive benefit. Non-value-added, required activities are activities that don't add value; however, these activities are required to the extent that people, processes, or systems need them. Lastly, non-value-added, not required activities don't have a benefit and therefore are waste and can be removed.

By shifting resources from non-value-added activities (required and not required), we have an opportunity to assess those activities that are necessary but don't provide real benefit and shift resources from essentially busywork (non-value-added activities) to value-added activities. The result is that we don't need to create a new business case for additional resources. Now, we may have a skills gap; however, it's reasonable to assume we can shift, reskill, or repurpose resources to meet our needs. Therefore, we don't require significant new headcount requests.

At this point, we've created a new value-management office and repurposed resources to focus on value-added activities. Our entire team is centered around developing and maturing our core capabilities. This brings us to identifying what those capabilities will be.

Business partners want to be empowered to and consume services based on performance. If that experience is good, more requests will come. Likewise, if that experience is negative, the requests will taper off.

Developing a value-management office service catalog helps make the services provided by the value-management office transparent to all business partners. This takes the guesswork out of what's provided as a standard offering and what's additional.

The concept of a service catalog in reference to a value-management office is especially important because of the new focus. Most project-management offices have a one-size-fits-all approach. The project is small, medium, or large and waterfall, hybrid, or agile and therefore has specific guidelines to follow. This means the project has a certain amount of overhead regardless of the scope or complexity of that project once the project's size and methodology are determined. The problem is this model doesn't scale efficiently.[114]

Companies are carving off projects of, say, <$500k or $1MM in size and classifying these as operational. Then these projects are being pushed back to business partners to manage internally. You probably think this won't work, because the business partners don't want them. In reality, we find that business leaders also identify these projects as small and would rather have full control over their delivery. This makes the business partners directly accountable for delivery, and it also frees up the project-management-office resources to spend more time on the value-management office; e.g., leading and managing the strategic organizational projects to ensure they maximize value as they deliver.[115]

The value-management office is focused only on strategic company initiatives. This means that services typically not provided by a project-management office are required to be provided as a service within a value-management office.

For example, advising executives on strategy, providing data-analytics services, or framing an enterprise-wide, organizational, change-management initiative needs to be supported. Each of these capabilities is crucial for the success of the value-management office. Often, the value-management officer or the business-information officer will hold positions on cross-functional teams; e.g., committees, councils, or governance teams. This alignment helps to ensure that adequate support is afforded to source, staff, and deliver the most critical strategic organizational projects.

The Value-management Office Pillars

Starting with a service-catalog framework for the new value-management office can be very helpful to communicate the services being offered. We'll begin by covering the outline of a sample service catalog and explain how it can be incorporated into a successful value-management office.[116]

The value-management-office service catalog has five primary pillars or capabilities:

1. Portfolio, program, and project management
2. Strategy and governance
3. Financial management
4. Planning and operations
5. People and resources

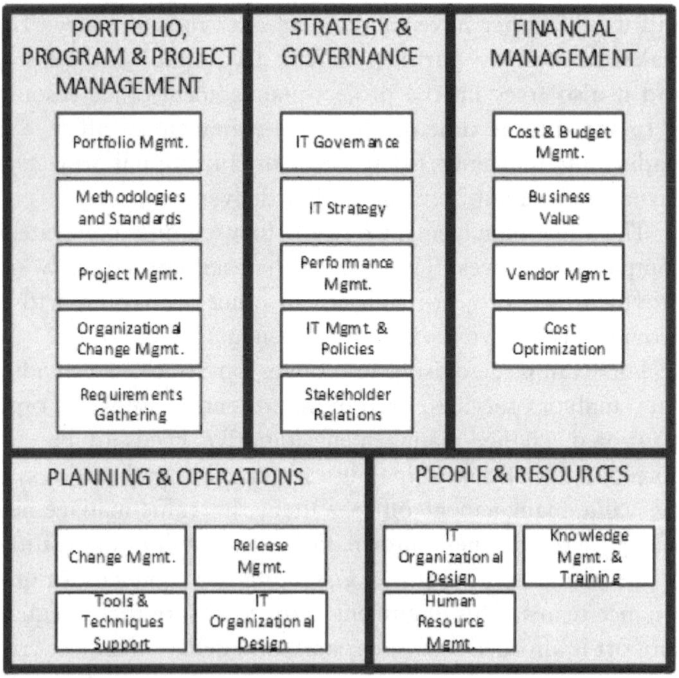

Portfolio, program, and project management covers delivery functions; e.g., portfolio management, project management, methodologies and standards, organizational-change management, and requirements gathering. Strategy and governance focus on the communication and escalation points for transparency; e.g., IT governance, IT strategy, performance management, IT management and policies, and stakeholder relations. Financial management wraps in vendor management to articulate value and control, monitor, and predict costs; e.g., cost and budget management, business value, vendor management, and cost optimization. Planning and operations are more the tactical, day-to-day business operations; e.g., change management, release management, tools and techniques support, and IT organizational design (think maintenance and operational teams that are often outsourced). Lastly, people and resources address how to maximize our resource effectiveness and efficiency; e.g., IT organizational design (department-wide operating model, not only operations), human-resource management, and knowledge management and training.

These five capabilities are further deconstructed into individualized services. Using this service catalog as a primary framework, we'll explore four of these services in more depth.

First, the IT governance service is part of the strategy and governance capability. This is an oversight function to control and monitor key information-technology-capability decisions. Historically, the concept of governance within a project-management office only referred to project governance—essentially, how information flowed within a project, program, or sprint. This definition has since been expanded within the framework of a value-focused organization. Governance encompasses any function required to deliver the strategic projects for the value-management office.[117]

Second, the IT strategy service is also part of the strategy and governance capability. This function provides advisory expertise to an executive leader to ensure that operational models properly connect strategy to execution. Traditionally, IT strategy had nothing to do with the project-management office. However, today, it has everything to do with the value-management office. This doesn't mean the value-management office directly controls the IT strategy. Yet, the value-management officer or business-information officer has a heavy influence in this space because, if technology strategy isn't tightly linked to business strategy, there will be delivery-value gaps within the strategic projects.[118]

Third, the vendor-management service is an essential part of the financial-management capability. This function provides managing, monitoring, and controlling of sourcing, pricing, contracting capabilities, and evaluation of performance by third parties. Procurement and category management (part of sourcing) often will actively own these vendor relationships. However, that ownership doesn't need to change in this model, although input into vendors' performance and adherence to performance measures do change. Vendors perform based on a contract; the value-management officer must influence the design, structure, and performance terms of the contracts required for strategic project execution.

Fourth, the human-resource management service is part of the people and resources capability. This function has ownership of the workforce-management functions of the team, department, and company. Everything of value produced is a product of people working together. Standard processes to engage talent must be in place to design an intelligent operating and staffing model that factors in talent development, growing a talent pipeline, and recruiting internally and externally for roles.

For consistency of delivery, as we review each service and apply it to a business context, we'll explain each one with the following four elements (an abridged scientific method) for simplicity:[119]

1. Service: the service being offered
2. Problem: the challenge this services addresses
3. Solution: how challenges are resolved or risk is mitigated by the service
4. Metrics: how we measure success

This format will ensure the models' application can be consumed by a wide range of readers, observers, and leaders.

The Power of IT Governance

The IT governance service validates alignment across the organization for multiple functions; e.g., technology, data, change, architecture, etc.

THE SERVICE

The IT governance service can comprise dozens of services. However, for the focus of this chapter, I've narrowed them down to eight essential governance services:

1. Enterprise
2. Enterprise project-management office
3. Technology policy and planning
4. Release management
5. Change
6. Data
7. Change-control board
8. Investment

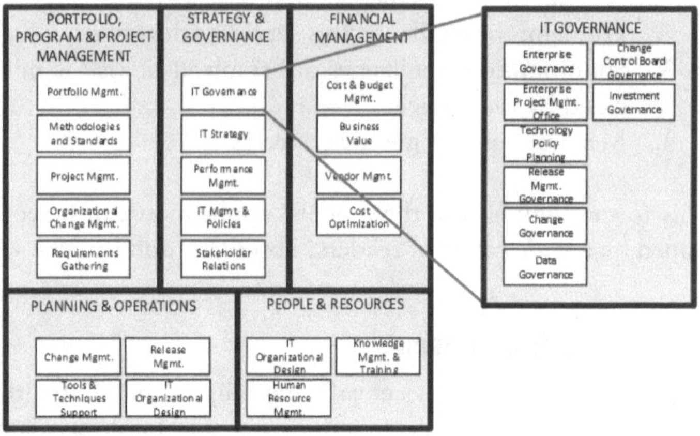

Enterprise governance is end-to-end governance of delivery (project-, program-, and executive-level governance). The enterprise project-management office is a function established to run a collection of portfolios to increase the project-management practice's maturity. Technology policy planning identifies new policy requirements and the impact of existing policies on the department. Release governance is the orchestration of delivery for the department. Change governance is the orchestration of the change-management process for the department. Data governance is the orchestration of the data-management process (data stewardship) for the department. The change-control-board governance is the orchestration of changes impacting delivery. Investment governance is the orchestration of investments across the portfolio for maximization of value for the department.[120]

THE PROBLEM

The IT governance service solves the problem of communication. In a larger organization, there are many functions, groups, and departments involved, and the value of the project or initiative can get lost. This erodes further when communications and decisions aren't clearly transmitted to all stakeholders contributing to creating value.

THE SOLUTION

The value-management office or officer is the single entity accountable for ensuring that full value is delivered for the strategic project or program. Of course, multiple groups contribute to the collective success, but a single entity needs to be accountable.

THE METRICS

IT governance is about communication and risk mitigation, but this can still be measured and quantified with KPIs. The seven metrics might include:

1. % of initial investment over total investments
2. % of projects linked to business growth investments
3. % delivery confidence
4. Number of monthly project changes
5. % of new projects from year-start baseline
6. $ spend by full-time equivalent resources
7. Number of services consumed per project

Metrics can cover services, processes, technology, delivery, or value. It's useful to track both leading and lagging indicators as part of the proactive governance model. After all, the objective is to gain knowledge and course-correct before a risk materializes.[121]

The Elements of Your IT Strategy

The IT strategy service connects the overall technology plan of objectives, principles, and tactics to the business strategy. This isn't a capability that commonly falls into the scope of a project-management office. However, this is squarely in the scope of the value-management office and an essential part of value realization.

THE SERVICE

IT strategy comprises multiple services (see below) and can shift based on the core capabilities of a company and the degree to which the value-management office leverages third parties for execution.[122]

1. Strategic-capability planning
2. Organizational operating model
3. Organizational role alignment
4. Organizational structures
5. Regulation identification
6. Maturity assessments
7. Capability and resource planning
8. Digital
9. Data
10. Cloud
11. IT operations
12. Enterprise architecture
13. Maintenance and operations
14. IT value
15. Technology innovation and ecosystems
16. External-interaction models
17. AI, ML, and AR

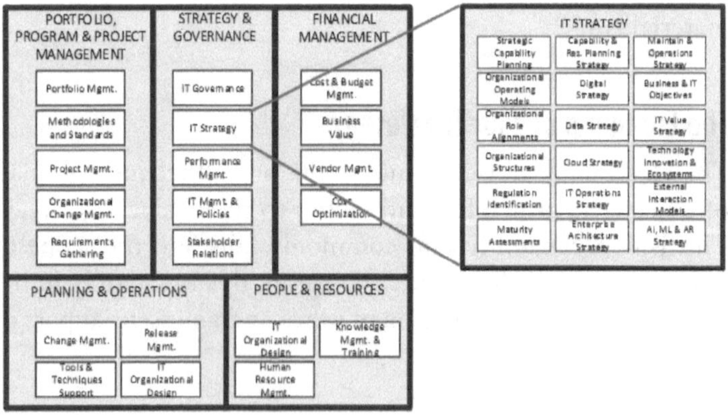

For simplicity, in this chapter, I've limited the IT strategy services.

Strategic-capability mapping is the alignment of business objectives to scientific capabilities to technology enablers (technology, systems, services). Organizational operating models is both an abstract and visual representation (model) of how an organization delivers value to its beneficiaries as well as how an organization actually runs itself. Organizational role alignment is a visual and narrative explanation of how roles converge to enable the target operating model. Organizational structures is the business model a department leverages to achieve business objectives. Regulation identification is the discovery of compliance or legal requirements and the analysis of their impact (SOPs, policy, legal, regulations, laws). Maturity assessments is the evaluation and measurement of point-in-time assessment of people, process, and technology maturity.

Capability-and-resource planning strategy is an approach ensuring that changes in an organization are aligned to the overarching strategic vision. A digital strategy is a plan and execution that uses digital resources to achieve one or more objectives. Data strategy is a plan and execution that uses data resources to perform one or more goals. Cloud strategy is a plan and execution that uses cloud resources to achieve one or more objectives. IT objectives are the development of the plan and measurable steps taken to achieve a goal. Enterprise architecture (EA) strategy is the practice of analyzing, designing, planning, and implementing enterprise analysis to execute business strategies successfully.[123]

Maintenance-and-operations strategy is the development and resulting approach taken to implement improvements in operations and maintenance during the entire asset lifecycle based upon competence, tools, and processes. IT value strategy is the definition, modeling, and realization of methods to capture value to present to your customers or partners. Technology innovation and ecosystems is the establishment of the variance players, influencers, and community members critical to achieving innovation adoption. External-interaction models is a visual illustration and narrative of how engagement between teams, departments, and organizations will be performed. AI, ML, and AR strategy is the development of an approach to incorporate these innovations in existing or new business processes for better business outcomes.

The visualization of these services for business partners helps create a mind shift to the value-management office that focus on outcomes rather than maintaining legacy, non-value-added processes.

THE PROBLEM

IT strategy connects operational tactics to business objectives and goals. The priority is to deliver the most strategic projects, which requires an understanding of the inputs that make up the total value model.

Rarely does value only originate from a project-management office or even a value-management office. Value maximization requires heavy, cross-department collaboration for the most impactful, valuable, and often complex initiatives the company is tackling.

THE SOLUTION

By linking data to cloud to digital to platforms, we get people talking, collaborating, and, ultimately developing a single integrated solution—not 10 separate but cool pieces of functionality. This approach helps prevent rework. The focus on the "right communication" the first time also leverages the reuse of business and technology components and modules.

THE METRICS

These metrics address how the strategy is being realized. Many of these metrics only have utility when presented in relation to a historical measure. Seven key metrics might include:

1. Number of strategic technology capabilities
2. % of cloud-enabled systems
3. $ of quarterly value meter growth
4. % of business capabilities delivered on the roadmap
5. % of operating-model staffing roles filled
6. % of technical capabilities at maturity
7. Number of technical capabilities with limitations

Metrics need to be fit-for-purpose. It's acceptable to adjust metrics based on your audience whether it be more technical or more business-focused.

Vendor Management

The vendor-management service enables maximum performance from third-party partners, suppliers, and vendors. Developing contracts and contractual performance measures ensures that delivery will be as expected. Complex delivery ecosystems often overlook vendor controls. Vendor control, management, and oversight is one of—if not the most critical—factors in delivery success. Additionally, if value measures are tied into vendor contracts, it helps align vendor and internal teams to partner to deliver and maximize value.[124]

THE SERVICE

The vendor-management service is within the IT governance capability and can comprise multiple, critical sub-services.

It's often assumed that leaders request support for all vendor-management services. It's worth the extra effort to clarify and specify which of these services they want to consume. This prevents offering services that are unwanted and frees up resources for more value-added activities. For our illustration, I'll highlight 18 potential vendor-management services.[125]

1. Supplier negotiations and management
2. Vendor governance

3. Contract management
4. Vendor-performance management
5. Vendor classifications
6. Vendor-solicitation lifecycle
7. New-employee management offering (NEMO) and coordination
8. Business-level agreements (BLAs)
9. Manage and maintain contracts
10. Manage contract and delivery expectations
11. Manage contractor administration
12. Vendor-provider selection
13. Contract submissions and changes
14. Work- or task-order management
15. Contract-dispute management
16. Vendor risk management
17. Vendor intellectual-property management
18. Contract negotiation

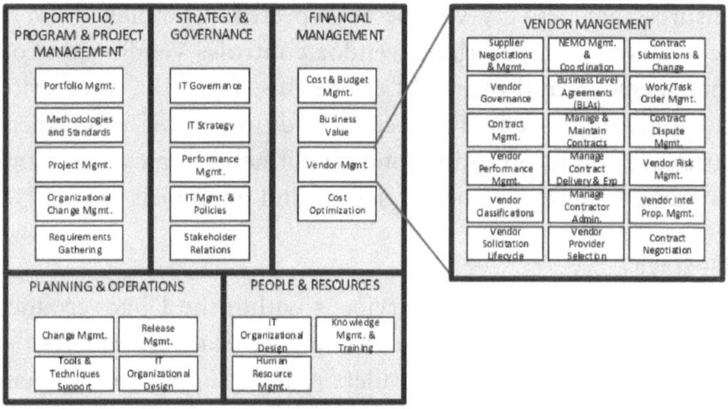

Supplier negotiations and management is the process of working directly with suppliers to ensure goods and services are delivered. Vendor governance is the oversight provided to vendors to ensure vendors meet contractual performance guarantees. Contract management is the active management of contracts through the pipeline (requested to be fully executed). Vendor-performance management is the management of vendor performance, similar to a vendor management office. Vendor classification is vendor-category management for performance optimizations (e.g., strategic partner, rate cards, etc.).[126]

A new-employee management offering and coordination is the process of working directly with NEMO to initiate and process new hires and monitor the status of existing contractors. Business-level agreements are the establishment and management of performance compared to benchmarks. Manage and maintain contracts is the proactive management of contracts for expiration and renewal. Manage contract-delivery and milestone expectations is the management of oversight to ensure vendors are meeting the contract terms and performance standards. Manage administration of contractors is the oversight and management of contractors including onboarding, training, timesheets, and contract renewals. Vendor-provider selection is the engagement, leadership, and participation in the vendor-selection processes.[127]

Contract submission and change control is the monitoring of contracts and initiates new change controls as required. Work- and task-order management is the development and management of the work/task orders for the department. Vendor-contract dispute management is the process of directly working with vendors when disagreements regarding the contract, future terms, or environmental factors impacting assumptions arise within the existing contract. Vendor risk management is the active management of vendor risk assumed by the department due to the distribution or lack of distribution of work across vendors. Vendor intellectual-property management is the process of managing the delivery of assets to the company to ensure compliance with internal standards for intellectual property. Contract negotiation is the process of drafting the language required to determine, establish, and measure vendor performance in scientific contracts.

THE PROBLEM

Vendors perform based on how they're measured. If high value is expected, the measures that support that outcome need to be established before the project is initiated. Otherwise, the vendor will perform to the contract, which may or may not result in the desired outcome.

THE SOLUTION

By closely partnering with procurement, category management, pricing, and legal value management, office executives can influence the direction of performance before course corrections are required.

THE METRICS

Useful metrics for vendors will be directly influenced by the contract, methodology of delivery, and history of each vendor. However, standard metrics should exist that span vendors over time to provide a historical baseline of vendor performance. This makes the identification of performance variance more visible. Here are seven relevant metrics that could be leveraged for vendor oversight:

1. Total annual spend by vendor YTD
2. % spend by business unit by the vendor
3. % vendor SLA non-conformance
4. % vendor on-time delivery milestones
5. Hours available, capacity dedicated
6. % of invoice with price variances from contract
7. Number of subcontractor violations by vendor type[128]

Human Resource Management

The human-resource-management service encompasses everything related to workforce management. This crucial service is the heart of all services that the value-management office provides. It's useful to validate how resources are flowing into and out of the value-management office to confirm that the right resources are placed into optimal roles.[129]

THE SERVICE

The human-resource service is broad, covering job training to counseling and development of resources and staff. This service is designed to optimize the most valued and costly asset to any enterprise—its people.

The intent of being specific with the variety of service offerings for human resources is to demonstrate that this function is often overlooked. It's assumed that, when a requisition is opened, a qualified resource will fill it. As we know, that isn't always the case. Building an intelligent operating model centered around continuous value delivery requires deliberate focus on managing human capital and designing operating models that result in change. Here, I've highlighted the most critical services. For sure, I could list many more that have value.[130]

1. Role description and responsibility development
2. Skills and competency inventory
3. Job training and education
4. Resources evaluation and benchmarking
5. Competency framework development
6. Recruiting and staffing
7. Contractor onboarding
8. Career ladders
9. Contract personnel management
10. Counseling and development
11. Policy and practices enforcement
12. Internship management

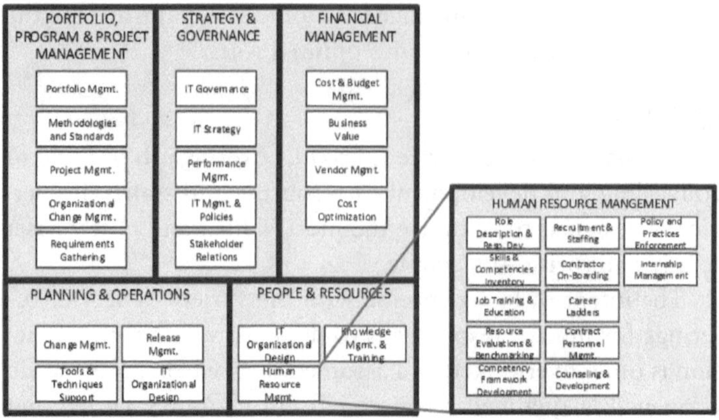

Role description and responsibility development is the design of roles and responsibilities to clarify and prevent role overlap and confusion. Skills and competency inventory is a repository of resources and their associated skills. Job training and education is the accessibility of tools and educational material necessary to do the job at hand. Resource evaluations and benchmarking is the process of evaluating FTEs and consultants for performance against a standard. Competency framework development is developing the framework used to define what "good" performance looks like.

Recruitment and staffing are the processes and steps taken to engage new talent for the organization and hedge against attrition and turnover. Contract onboarding is the process of taking a contractor unfamiliar with the company and guiding them through the setup and onboarding process. Career ladders are a narrative and visual path for professional-development progress for an individual. Contract personnel management is contractor oversight and management to ensure performance meets the set standard and takes action steps if that doesn't occur. Counseling and development is the process of coaching FTEs to ensure departmental-role, individual, and development goals are achieved.

Policy and practice enforcement is the reinforcement of policies and departmental standards (best practices) to ensure compliance. Internship management is the process to request, engage, and guide a new intern into the department.

THE PROBLEM

Human-resource management is a collaborative service. Building a rapport with your human-resource partner is useful when developing and executing the human-capital operating model. Failure to establish the necessary buy-in will result in canceled or delayed requisitions or, at worst, a negative impact on the relationship with your human-resource partner.

THE SOLUTION

The development of an operating model must take into consideration the resources that need to be removed, reskilled, or moved up. It's virtually impossible to hire new personnel with all the required resources needed to deliver value for transformative and complex organizational strategic projects. As a result, strategic planning that includes human-capital management is a great option to provide human-resource management partners with a vision for the future and how they fit into that value-realization model.[131]

THE METRICS

The metrics for human-capital management tend to be more stable than in other areas. As a result, it's easier to forecast standards and leverage industry benchmarks for attrition and turnover, for example. These seven metrics are a great start to getting control of this service performance:

1. Number of voluntary turnovers
2. % of open roles filled internally
3. % of defined positions with an identified successor
4. Total amount spent on support of employees
5. Total headcount and total full-time-equivalent (FTE) headcount

6. Revenue minus purchased services per FTE (human-capital profit index)
7. Total pay and benefits costs + pay costs for contingent workforce + costs of absenteeism + costs of turnover (human-capital cost factor)

When using these or the numerous additional human-capital metrics, value-management-office leaders gain better insight into how staffing and resources impact financials and affect value-based outcomes.

Research Propositions

The following propositions and potential applications are based on my research in methods to maximize the value-management office's impact in accelerating value enterprise-wide.

Three propositions (P1, P2, and P3) are offered that have the potential to introduce impactful approaches to advance enterprise value-management capabilities.

PROPOSITION 1: DRIVING INNOVATION THROUGH COOPETITION STRATEGIES

Recent innovations have disrupted how PMOs function. Cloud solutions, digitization, artificial intelligence, blockchain, voice-driven technology, advanced robotics, and gnomonic advances have all impacted how PMOs operate. People's knowledge and understanding of how to solve problems have evolved. Legacy PMO processes no long support emerging technologies. PMOs must recognize a change has occurred.[132]

The necessity to swiftly change, support new technologies, and continually learn is too much for many companies to handle in isolation. However, with coopetition strategies and partnering with your customary competitors, companies learn to adapt and elevate their game of play.

Sharing how innovative tools, products, and services are utilized doesn't disclose confidential information but enlightens efficiencies that weren't in place just days prior. Strategic companies are adding innovative capabilities through coopetition and taking smart risks together.

PROPOSITION 2: AGILE APPROACHES TO DELIVERING VALUE TO THE ENTERPRISE

Strategic leaders accountable for value-based delivery are finding that the skills they've developed in the PMO are no longer effective. Waterfall thinking with minimum viable products and project-management software isn't getting the job done. This approach is simply not agile enough and doesn't support the degree of change that's taking place within our business ecosystems.

Agile approaches, managing change, intelligence dashboards, and cross-functionally-trained leadership are rising to the top of committee and council meetings. PMOs are absorbing more strategically aligned projects. These projects demand that teams collaborate much more closely than was the case with legacy-modernization efforts of the past.

New metrics, insights, and analytics reassess how we determine success and manage risk. How quickly a new project or initiative can be added, prioritized, staffed, and started is a new strategic-agility measure.[133,134]

PROPOSITION 3: EMPOWER EMPLOYEES TO FUEL INNOVATION

Innovation is required to compete today, and it will be a requirement to be competitive in the future. Freeing up 10 or 50 resources to focus on innovative stuff won't rocket your company into the future. The lift required is too great.

Innovation, at its heart, is about people—more specifically, how you're managing them. It's less about performance reviews, but that's part of it. Enterprise-resource planning is the missing key element to achieve game-changing innovation.

How do you achieve breakthrough innovations if your employees are busy supporting legacy systems and following inefficient processes? Shifting an organization from a traditional project-management office that provides limited organizational value to a high-functioning, strategic, value-management office requires the right people in the right roles. We can't have folks working on projects or programs or doing work that doesn't directly contribute to the bottom line. In short, if work doesn't enable, directly or indirectly, a strategic objective, we need to find a way to remove or stop the non-value-added activity.

The ability to adapt and change is directly related to the maturity of that same organization's enterprise-resource-management capability.

A leader needs visibility into resource forecasting to determine how much they're able to adapt and change, and this is prioritized based on the resources in place today. It doesn't sound new age, but it is.

Limitations and Direction of Future Work

First, this chapter contributes to the expansive body of knowledge focused directly on project-management-office development and maturity. It also attempts to shift focus to the much narrower research areas concentrated on the progressive value-management office. The industry will soon shift from the maintenance and operational, tactically focused PMO to a value-management office with strategic and enterprise-wide focus. This shift will swing new responsibilities into the value-management office. Five of these new responsibilities will include:

1. Strategic initiatives
2. Agility
3. Automation
4. Collaboration
5. Data and analytics

Strategic initiatives will consume the value-management office, and projects of lesser importance will be pushed back to business partners to deliver at a lower risk. Agility will be measured not in speed of project execution but in the time required to shift focus from project A to project B. Automation, in the form of artificial intelligence, will combine with machine learning to create intelligence platforms that drive insights to the portfolio, program, and project teams without them having to ask. Collaboration necessitates that most projects—due to their strategic nature—will require more interaction from more players. Instead of two or three departments, some initiatives will require nearly all departments to be involved. This will alter how communication is done enterprise-wide and how interdepartmental interactions are managed and governed. Data and analytics will provide an answer to questions we didn't know needed to be asked. When compiled with data technologies, statistical algorithms will provide visualizations that are actionable and help drive value realization through the enterprise.[135]

The PMO is being disrupted. Leaders who don't learn are being replaced with leaders that do. New questions are being asked of PMO leaders:

1. How can AI advance or accelerate the portfolio?
2. What percentage of strategic projects could be shortened with improved data-quality and data-cleansing tools?
3. Which functions within the PMO can be automated using robotic process automation (RPA) technologies?

The requirements for how to be strategic are changing. Skills that made portfolio executives effective at driving a strategic portfolio 10 years ago aren't the skills required today to lead that same portfolio. The challenges of the new innovative portfolio executive are real.

1. Do you understand the relationship among artificial intelligence, intelligent agents, and machine learning?

If not, how are you going to evaluate the impact on your strategic initiatives?

2. Are you familiar with Drake, TIBCO Clarity, Cloudingo, Reifier? How will you understand the limits of what can or can't be done to transform, clean, and massage data required for 90% of your portfolio?
3. Don't have experience with RPA? Then how can you accelerate digital approvals, push project variances, or automate forecasts based on historical data using RPA?

CEOs and CIOs that are staffing executives to lead company transformation into value-management offices appreciate the new skills required today to be successful.

The executive leaders accountable for the value-management office and championing strategic initiatives don't need to be experts in data and innovation. They do, however, need to understand the basic concepts I've outlined, and many more, to be effective and compete in this dynamically changing and agile world.[136]

Do You Have the Right Leader in Place?

Project-management offices are either being dissolved or transformed into value-management offices that are refocused on maximizing organizational value.

This might appear to be a branding stunt. That couldn't be farther from the truth. This new paradigm of thinking is transformational and requires every resource in the organization to think differently about what they're doing and, more specifically, how they're doing it. Every employee, every day should be asking themselves, "Am I adding value today?" This changes the discussion from, "Let's do something"—implying almost anything is acceptable—to "Let's do something of value."

To provide value, stakeholders—your business partners, peers, and colleagues—need to understand what services your organization provides to the enterprise. The project-management office of yesterday was focused on defining processes, setting up cohort discussions, and reevaluating templates that were rarely used and rarely used correctly.

It didn't work. No one bought it. If we shift our energy from a cost-center to a value-center mentality with quantified impacts, now we have something. This new value-management office is impactful and can be quantified and shared. Now, stakeholders begin to understand the services we provide and how they're provided. We're empowering our business partners.

If value is added, they consume more; if value is removed, they consume less. It's a simple concept. This dismantles the two-way mirrors in which no one understands what the resources inside the project-management office are working on. We know they're in there working on something, but what they're working on is unknown.

This hyper-transparency builds trust. This trust is contagious throughout the organization, and resources internalize what's important to the organization not only to create value but to sustain it. How many projects a project manager delivers no longer matters. What does matter is the impact or value realized from a single project.

The specifics of what's included in the service catalog really isn't significant. What's relevant is that the service catalog explains what service you provide and how it's provided.

Value-management offices spend more time on advising, coaching, and collaborating to develop strategic plans and visions than monitoring vendor invoices. Those detailed activities are important, but there are tools—automation, RPA, and the like—that perform these functions better than humans.

Position your team to have the best chance to maximize the value realized within your organization. Reflect for a minute on whether you have the right people in the right role to transform your PMO into a value-management office. Consider these questions as you envision the future:

1. Do my leaders have the innovation experience to understand how the impact of technology can maximize value?
2. Is the best operational and engagement model in place to align roles and responsibilities to the capabilities that will drive our strategic projects?
3. Are we using data and analytics as a strategic capability within the PMO today?

If the answer to any of these questions is no, you can only improve from here. What's important is to acknowledge that you don't have the right leaders in place to champion that transformation.

Project-management offices overflowing with templates and process red-tape will soon be a thing of the past. Value-management offices will emerge to discover, capture, and maximize organizational value by focusing on the most strategic enterprise initiatives.

Value-management offices—led by visionary executives—will transform innovative organizations into value factories.

I believe that executives who are accountable for value maximization understand that organizational leadership changes are required. Bold executives, with the confidence to make strategic hires, will ultimately maximize organizational value through the successful execution of strategic initiatives.

CITING THE BOOK

APA

Nichol, P. B. (2021). Leading with Value: How to Effectively Communicate Business Value (1st ed.). OROCA Innovations.

CHICAGO

Nichol, Peter B. Leading with Value: How to Effectively Communicate Business Value (1st ed.). Newington, CT USA: OROCA Innovations, 2021.

ELSEVIER HARVARD

Nichol, P 2021, Leading with Value: How to Effectively Communicate Business Value, 1st ed. OROCA Innovations, Connecticut, USA.

IEEE

P. Nichol, Leading with Value: How to Effectively Communicate Business Value, 1st ed. Newington, CT USA: OROCA Innovations, 2021.

ENDNOTES

CHAPTER 1

1. Letavec, C. (2006). *The Program Management Office: Establishing, Managing And Growing the Value of a PMO* (Illustrated Edition). J. Ross Publishing.
2. Ferenc, S. (2011). *Analyzing Economic Value Creation in Terms of the Infocommunications Investments of Domestic Enterprises*. Hungarian Academy of Sciences.
3. Aranyossy, M. (2014). How to Measure Business Value of Information Systems? Practical Implications of a Literature Review. *International Journal of Advanced Research in Computer and Communication Engineering*, 3(11), 7.
4. Value Quantification Framework® is a value management methodology developed by Peter B. Nichol and registered with the U.S. Copyright Office.
5. Baldwin, P. (2020). *Profit Peak for Entrepreneurs: A Practical, Field-tested Guide to Maximizing Profit and Optimizing Cash Flow* (M. Carlisle, Ed.). Independently published.
6. Mascarella, G. (2005). *Rapid Economic Justification: A Step-by-Step Guide to Optimizing IT Investments that Forge Alliances Between IT and Business* (p. 146). Microsoft Corporation.
7. Taylor, P., & Mead, R. (2015). *Delivering Successful PMOs: How to Design and Deliver the Best Project Management Office for your Business* (1st Edition). Routledge.
8. Tapping, D., Luyster, T., & Shuker, T. (2002). *Value Stream Management: Eight Steps to Planning, Mapping, and Sustaining Lean Improvements* (PAP/CDR Edition). Productivity Press.
9. Tapping, D., & Shuker, T. (2003). Value *Stream Management for the Lean Office: Eight Steps to Planning, Map-*

ping, & Sustaining Lean Improvements in Administrative Areas (PAP/CDR Edition). Productivity Press.

10. Taylor, P. (2016). *Leading Successful PMOs: How to Build the Best Project Management Office for Your Business* (1st Edition). Routledge.
11. FitzGerald, M. (2017). *Customer-Centric Cost Reduction: How to invest and improve profits without sacrificing your customers*. SBVV.
12. Peters, J. (2019). *Agile Project Management: The Professional Step-by-Step Guide for Beginners to Deeply Understand Agile Principles From Beginning to End, Developing Agile Leadership and Improving Soft Skills*. Independently published.
13. Robison, D. J. A. (2018). *Project Math: Tools and Techniques for Project Managers, Agile Coaches and Scrum Masters, Project Sponsors and Business Analysts, Project Management Offices, Team Members, and Engaged Stakeholders*. Independently published.
14. Tierney, N. W. (2017). *Value Management in Healthcare: How to Establish a Value Management Office to Support Value-Based Outcomes in Healthcare* (1st Edition). Productivity Press.
15. Crawford, J. K. (2010). *The Strategic Project Office* (2nd Edition). CRC Press.
16. Duggal, J. (2018). *The DNA of Strategy Execution: Next Generation Project Management and PMO* (1st Edition). Wiley.
17. Hill, G. M. (2007). *The Complete Project Management Office Handbook*, Second Edition (2nd Edition). Auerbach Publications.
18. Venkataraman, R. R., & Pinto, J. K. (2008). *Cost and Value Management in Projects* (1st Edition). Wiley.
19. Slot, R. G. (2010). *A method for valuing architecture-based business transformation and measuring the value of solutions architecture*. University of Amsterdam, 214.

20. Sasvári, P. (2012). Adaptable Techniques for Making IT-Related Investment Decisions. *International Journal of Advanced Research in Computer Science and Software Engineering*, 2(2).
21. Ibid.
22. Slot, *A method for valuing architecture-based business transformation and measuring the value of solutions architecture.*
23. Ibid.
24. Kerzner, H. (2017). *Project Management: A Systems Approach to Planning, Scheduling, and Controlling* (12th Edition). Wiley.
25. Begovac, D., Fagerström, S., & Nobel, M. (2005). *What aspects are important in an ex-post investment evaluation?* (p. 70). IT University of Göteborg.
26. Kerzner, H. (2017). *Project Management Metrics, KPIs, and Dashboards: A Guide to Measuring and Monitoring Project Performance* (3rd Edition). Wiley.
27. Kerzner, *Project Management: A Systems Approach to Planning, Scheduling, and Controlling.*
28. Slot, *A method for valuing architecture-based business transformation and measuring the value of solutions architecture.*
29. Begovac, *What aspects are important in an ex-post investment evaluation?.*
30. McManus, PhD, H. L. (2005). *Product Development Value Stream Mapping (PDVSM) Manual* (p. 116). Massachusetts Institute of Technology.
31. Kogon, K., Blakemore, S., & Wood, J. (2015). *Franklin-Covey Project Management for The Unofficial Project Manager,* paperback (1st Edition). FranklinCovey.
32. Aubry, M., & Hobbs, B. (2010). *The Project Management Office* (Illustrated Edition). Project Management Institute.

CHAPTER 2

1. er, E. R. (2018). *Leading Exponential Change: Go beyond Agile and Scrum to run even better business transformations*. Innova1st Publishing.
2. Feliciano, Dan (2020). Personal interview [Personal discussion].
3. Ibid.
4. Bevans, R. (2020, August 7). One-way ANOVA | When and How to Use It (With Examples). Scribbr. https://www.scribbr.com/statistics/one-way-anova/
5. Six Sigma Guide. (2017, August 15). One- and Two-Sample Proportion Hypothesis Tests. Six Sigma Study Guide. https://sixsigmastudyguide.com/one-and-two-sample-proportion-hypothesis-tests/
6. Statistics Solutions. (2020). Conduct and Interpret a One-Sample T-Test. Statistics Solutions. https://www.statisticssolutions.com/conduct-interpret-one-sample-t-test/
7. Liker, J. (2004). The Toyota Way: 14 Management Principles from the World's Greatest Manufacturer (1st Edition). McGraw-Hill Education.
8. Kerzner, H. (2017). Project Management Case Studies (5th Edition). Wiley.
9. Six-Sigma-Material. (Unknown). I-MR, Individuals—Moving Range Charts. https://www.six-sigma-material.com/I-MR.html
10. Wisner, J. D., Tan, K.-C., & Leong, G. K. (2018). Principles of Supply Chain Management: A Balanced Approach (5th Edition). Cengage Learning.
11. George, M. L., Maxey, J., Rowlands, D., & Price, M. (2004b). The Lean Six Sigma Pocket Toolbook: A Quick Reference Guide to 100 Tools for Improving Quality and Speed (1st Edition). McGraw-Hill.
12. SAS Institute Inc. (2014, February). One Sample T-Test and CI. https://www.stat.purdue.edu/~lfindsen/stat511/04_one_sample_t-test_and_ci.pdf

13. Zaiontz, C. (2020, August 21). Real Statistics Resource Pack. http://www.real-statistics.com/free-download/real-statistics-resource-pack/
14. GOAL/QPC, Carleton, S., & Academy, S. S. (2016). The Black Belt Memory Jogger (2nd Edition). Goal/QPC.
15. Laerd Statistics. (2020). One-Way ANOVA - How to report the significance results, homogeneity of variance and running post-hoc tests. https://statistics.laerd.com/statistical-guides/one-way-anova-statistical-guide-4.php
16. Brenig-Jones, M., & Dowdall, J. (2018). Lean Six Sigma For Leaders: A practical guide for leaders to transform the way they run their organization (1st Edition). Wiley.
17. SPSS Tutorials. (2020). ANOVA (Analysis of Variance)—Super Simple Introduction. https://www.spss-tutorials.com/anova-what-is-it/
18. Yeager, K. (2020). LibGuides: SPSS Tutorials: One-Way ANOVA. https://libguides.library.kent.edu/SPSS/OneWayANOVA
19. Bass, I. (2007). Six Sigma Statistics with EXCEL and MINITAB (1st Edition). McGraw-Hill Education.
20. Khan, M. R. M. (2016a). Six Sigma Statistics using Minitab17.: Green Belt Edition. (1st Edition). CreateSpace Independent Publishing Platform.
21. Minitab, LLC. (2013, June 19). How to Create and Read an I-MR Control Chart. https://blog.minitab.com/blog/understanding-statistics/how-create-and-read-an-i-mr-control-chart
22. Mohiuddin, G. (2011). IMPLEMENTING SIX SIGMA TO REDUCE REJECTIONS IN A MANUFACTURING COMPANY: Concepts and techniques to reduce rejections and reworks to improve customer service. LAP LAMBERT Academic Publishing.
23. Wild, C. J., & Seber, G. A. F. (2020). Chapter 13 of Chance Encounters: Control Charts. https://www.spss-tutorials.com/anova-what-is-it/

24. Basu, R. (2008). Implementing Six Sigma and Lean: A practical guide to tools and techniques (1st Edition). Butterworth-Heinemann.
25. Kotter, J. P. (2012). Leading Change, With a New Preface by the Author (1R Edition). Harvard Business Review Press.
26. Feliciano, Dan (2020). Personal interview [Personal discussion].
27. Pyzdek, T., & Keller, P. (2018). The Six Sigma Handbook, 5E (5th Edition). McGraw-Hill Education.
28. Toyoda, R. (2020). Project Management: The Ultimate Guide to Help You Master and Innovate Projects with Lean Thinking, Including How to Dominate Agile, Scrum, Kanban, and Six Sigma. Independently published.
29. Laerd Statistics. (Unknown). Two-way ANOVA in SPSS Statistics. https://secure.livechatinc.com/
30. George, M. L., Maxey, J., Rowlands, D., & Price, M. (2004a). The Lean Six Sigma Pocket Toolbook: A Quick Reference Guide to 100 Tools for Improving Quality and Speed (1st Edition). McGraw-Hill.
31. Martinez, David (2020). Personal interview [Personal discussion].
32. Ibid.
33. Pande, P. S., Neuman, R. P., & Cavanagh, R. R. (2002). The Six Sigma Way Team Fieldbook: An Implementation Guide for Process Improvement Teams (1st Edition). McGraw-Hill Education.
34. SAS Institute Inc. (Unknown). Multiple Linear Regression. https://www.jmp.com/en_us/statistics-knowledge-portal/what-is-multiple-regression.html
35. Brook, Q. (2020). Lean Six Sigma and Minitab (6th Edition): The Complete Toolbox Guide for Business Improvement (6th Edition Revised). OPEX Resources Ltd.
36. Productivity-Quality Systems, Inc. (Unknown). p-chart (What is it? When is it used?). https://www.pqsystems.com/qualityadvisor/DataAnalysisTools/p_chart.php

37. Gardner, K. M. (2013, January 1). Successfully Implementing Lean Six Sigma: The Lean Six Sigma Deployment Roadmap.
38. Breyfogle III, F. W. (2003). Implementing Six Sigma: Smarter Solutions Using Statistical Methods (2nd Edition). John Wiley & Sons.
39. Defeo, J. (2016). Juran's Quality Handbook: The Complete Guide to Performance Excellence, Seventh Edition (7th Edition). McGraw-Hill Education.
40. The Council for Six Sigma Certification. (2018). Six Sigma: A Complete Step-by-Step Guide: A Complete Training & Reference Guide for White Belts, Yellow Belts, Green Belts, and Black Belts. The Council for Six Sigma Certification.

CHAPTER 3

1. Nir, M. (2014). *The Agile PMO: Leading the Effective, Value Driven, Project Management Office* (C. Nir, Ed.; 4th Edition). CreateSpace Independent Publishing Platform.
2. Tapping, D., & Shuker, T. (2003). Va*lue Stream Management for the Lean Office: Eight Steps to Planning, Mapping, & Sustaining Lean Improvements in Administrative Areas* (PAP/CDR Edition). Productivity Press.
3. Kerzner, H. (2017). *Project Management Metrics, KPIs, and Dashboards: A Guide to Measuring and Monitoring Project Performance* (3rd Edition). Wiley.
4. Chatzipanos, D. P., & Wu, D. T. (2018). *Implementing Project Portfolio Management*. Project Management Institute.
5. Chincarini, L. B., & Kim, D. (2006). *Quantitative Equity Portfolio Management: An Active Approach to Portfolio Construction and Management* (1st Edition). McGraw-Hill Education.

6. Taylor, P. (2016). *Leading Successful PMOs: How to Build the Best Project Management Office for Your Business* (1st Edition). Routledge.
7. Basu, R. (2008). *Implementing Six Sigma and Lean: A practical guide to tools and techniques* (1st Edition). Butterworth-Heinemann.
8. Aubry, M., & Hobbs, B. (2010). *The Project Management Office* (Illustrated Edition). Project Management Institute.
9. Husser, P. (2017). *The High-Impact PMO: How Agile Project Management Offices Deliver Value in a Complex World* (1st Edition). CreateSpace Independent Publishing Platform.
10. Defeo, J. (2016). *Juran's Quality Handbook: The Complete Guide to Performance Excellence, Seventh Edition* (7th Edition). McGraw-Hill Education.
11. Womack, J. P., Jones, D. T., & Roos, D. (2007). *The Machine That Changed the World: The Story of Lean Production-- Toyota's Secret Weapon in the Global Car Wars That Is Now Revolutionizing World Industry* (Reprint edition). Free Press.
12. Brenig-Jones, M., & Dowdall, J. (2018). *Lean Six Sigma For Leaders: A practical guide for leaders to transform the way they run their organization* (1st Edition). Wiley.
13. Dennis, P. (2015). *Lean Production Simplified*, Third Edition (3 edition). Routledge.
14. Robison, D. J. A. (2018). *Project Math: Tools and Techniques for Project Managers, Agile Coaches and Scrum Masters, Project Sponsors and Business Analysts, Project Management Offices, Team Members, and Engaged Stakeholders*. Independently published.
15. Toyoda, R. (2020). *Project Management: The Ultimate Guide to Help You Master and Innovate Projects with Lean Thinking, Including How to Dominate Agile, Scrum, Kanban, and Six Sigma*. Independently published.

16. Martin, K., & Osterling, M. (2013). *Value Stream Mapping: How to Visualize Work and Align Leadership for Organizational Transformation* (1 edition). McGraw-Hill Education.
17. Bolles, D. L., & Hubbard, D. G. (2014). *The Power of Enterprise PMOs and Enterprise-Wide Project Management: Introducing a Business Management Model Integrating and Harmonizing Operations Business Management and Project Business Management.* PBMconcepts.
18. Venkataraman, R. R., & Pinto, J. K. (2008). *Cost and Value Management in Projects* (1st Edition). Wiley.
19. Hofstede, G., Hofstede, G. J., & Minkov, M. (2010). *Cultures and Organizations: Software of the Mind, Third Edition* (3rd Edition). McGraw-Hill Education.

CHAPTER 4

1. Govindarajan, V. (2010). *Innovation is Not Creativity.* Harvard Business Review. https://hbr.org/2010/08/innovation-is-not-creativity
2. Nichol, P. B. (2018). *Redefine the future of your organization with a competency framework.* https://www.cio.com/article/3254697/redefine-the-future-of-your-organization-with-a-competency-framework.html
3. Mair, A. (1999). *Learning from Japan? Interpretations of Honda Motors.* Strategic Management Theorists.
4. Kerzner, H. (2017). *Project Management: A Systems Approach to Planning, Scheduling, and Controlling* (12th Edition). Wiley.
5. Kerzner, H. (2017). *Project Management: A Systems Approach to Planning, Scheduling, and Controlling* (12th Edition). Wiley.
6. Kerzner, H. (2017). *Project Management Metrics, KPIs, and Dashboards: A Guide to Measuring and Monitoring Project Performance* (3rd Edition). Wiley.
7. Ibid

8. Gartner. (2016). *Information Life Cycle Management (ILM)—Gartner IT Glossary.* http://www.gartner.com/it-glossary/information-life-cycle-management-ilm
9. Kerzner, H. (2017). *Project Management Metrics, KPIs, and Dashboards: A Guide to Measuring and Monitoring Project Performance* (3rd Edition). Wiley.
10. Crawford, J. K. (2014). *Project Management Maturity Model* (3rd Edition). Auerbach Publications.
11. Ibid.
12. Ibid.
13. Toyoda, R. (2020). *Project Management: The Ultimate Guide to Help You Master and Innovate Projects with Lean Thinking, Including How to Dominate Agile, Scrum, Kanban, and Six Sigma.* Independently published.
14. Womack, J. P., & Jones, D. T. (2003). *Lean Thinking: Banish Waste and Create Wealth in Your Corporation, Revised and Updated* (2nd edition). Free Press.
15. Crawford, J. K. (2014). *Project Management Maturity Model* (3rd Edition). Auerbach Publications.
16. Nichol, P. B. (2019, October 29). *First, define "good" to scale your organization.* CIO. https://www.cio.com/article/3448545/first-define-good-to-scale-your-organization.html
17. Moore, K. (2011). *Porter or Mintzberg: Whose View of Strategy Is the Most Relevant Today?* Forbes.
18. Robison, D. J. A. (2018). *Project Math: Tools and Techniques for Project Managers, Agile Coaches and Scrum Masters, Project Sponsors and Business Analysts, Project Management Offices, Team Members, and Engaged Stakeholders*. Independently published.
19. Moreira, M. E. (2017). *The Agile Enterprise: Building and Running Agile Organizations* (1st ed. Edition). Apress.

20. Nichol, P. B. (2017). *Learning Intelligence: Expand Thinking, Absorb Alternatives and Unlock Possibilities* (1st ed.). OROCA Innovation Labs.
21. Nichol, P. B. (2020). *Quantifying the value of your enterprise data-management initiative.* https://leadersneedpancakes.com/quantifying-the-value-of-your-enterprise-data-management-initiative-2/

CHAPTER 5

1. Husser, P. (2017). *The High-Impact PMO: How Agile Project Management Offices Deliver Value in a Complex World* (1st Edition). CreateSpace Independent Publishing Platform.
2. Nir, M. (2014). *The Agile PMO: Leading the Effective, Value Driven, Project Management Office* (C. Nir, Ed.; 4th Edition). CreateSpace Independent Publishing Platform.
3. Gallacher, L., & Morris, H. (2012). *ITIL Foundation Exam Study Guide* (1st Edition). Sybex.
4. McLean, Dan. (2014). *Catalogs, Services and Portfolios: An Itsm Success Story*. ITGP.
5. Wood, M. (2016). *Mastering ServiceNow*—Second Edition (2nd Revised edition). Packt Publishing.
6. Moreira, M. E. (2017). *The Agile Enterprise: Building and Running Agile Organizations* (1st ed. Edition). Apress.
7. Bolles, D. L., & Hubbard, D. G. (2014). *The Power of Enterprise PMOs and Enterprise-Wide Project Management: Introducing a Business Management Model Integrating and Harmonizing Operations Business Management and Project Business Management.* PBMconcepts.
8. Hofstede, G., Hofstede, G. J., & Minkov, M. (2010). *Cultures and Organizations: Software of the Mind*, Third Edition (3rd Edition). McGraw-Hill Education.
9. Chincarini, L. B., & Kim, D. (2006). *Quantitative Equity Portfolio Management: An Active Approach to*

Portfolio Construction and Management (1st Edition). McGraw-Hill Education.

10. Axelos. (2020). *ITIL foundation.* IT Governance Publishing.
11. Duggal, J. (2018). *The DNA of Strategy Execution: Next Generation Project Management and PMO* (1st Edition). Wiley.
12. Chatzipanos, D. P., & Wu, D. T. (2018). *Implementing Project Portfolio Management.* Project Management Institute.
13. Axelos. (2020). *ITIL 4 Create, deliver and support.* TSO.
14. McLean, Daniel. (2012). *No One of Us is as Strong as All of Us: Services, Catalogs and Portfolios* (I. G. Publishing, Ed.). IT Governance Publishing.
15. Josephs, A., & Rubenstein, B. (2018). *Risk Up Front: Managing Projects in a Complex World.* Lioncrest Publishing.
16. Reilly, F. K., Brown, K. C., & Leeds, S. J. (2018). *Investment Analysis and Portfolio Management* (11th Edition). Cengage Learning.
17. Nair, S. (2020). *The Service Desk Handbook: A guide to service desk implementation, management and support.* ITGP.
18. Spivak, E. (2019). *PMO Governance: Practical Strategies to Govern Portfolio, Program, and Project Delivery.* FriesenPress.
19. Steinberg, R. A. (2007). *Servicing ITIL: A Handbook of IT Services for ITIL Managers and Practitioners.* Trafford Publishing.
20. Van Haren Publishing (Ed.). (2019). *ITIL®4 – A Pocket Guide.* Van Haren Publishing.
21. Stewart, S. D., Piros, C. D., & Heisler, J. C. (2019). *Portfolio Management: Theory and Practice* (1st Edition). Wiley.

22. Campbell, A. (2020). *Agile and Scrum: Complete Guide. What is Agile and What is Scrum?* Independently published.
23. Osborn, P. (2014). *Succeeding with Agile Teams*: Volume 2. CreateSpace Independent Publishing Platform.
24. Bühler, E. R. (2018). *Leading Exponential Change: Go beyond Agile and Scrum to run even better business transformations*. Innova1st Publishing.

INDEX

From OROCA Innovations: Author Peter B. Nichol, a 4x author, certified Master Business Relationship Manager (MBRM®) by the BRM Institute and a MIT Sloan School of Management-recognized CIO has finally captured how the most innovative leaders connect strategy to execution—in his most recent book entitled, *Think Lead Disrupt!*

"*Think Lead Disrupt* is a fantastic book for leaders that want to develop innovative ideas that can be translated into world-class business products, services, and interactions. Nichol's voice as a CIO and global thought leader is simple to understand and easy to apply. This is a must read for innovators leading teams, departments, or organizations that must be on the cutting edge!"

AARON S. BARNES
CEO AND CO-FOUNDER
BUSINESS RELATIONSHIP MANAGEMENT (BRM) INSTITUTE

From OROCA Innovations: Author Peter B. Nichol, an MIT Sloan School of Management-recognized CIO and award-winning innovation leader, explores the for-gotten world of learning—the ability to learn—in the book entitled, *Learning Intelligence.*

"Peter dives into unexplored, valuable territory in *Learning Intelligence*. A must-read for any leader wanting to compete in the innovation-powered landscape of today."

MARSHALL GOLDSMITH,
THE AUTHOR OF THE #1 NEW YORK TIMES BESTSELLER,
TRIGGERS.

Recently from OROCA Innovations: Author Peter B. Nichol, an MIT Sloan School of Management-recognized CIO and award-winning innovation lead-er, explores how blockchain transform the patient experience—in the book entitled, *The Power of Block-chain for Healthcare.*

"*The Power of Blockchain for Healthcare* provides a compelling argument for the transformative impact that blockchain will have on the patience experience. Nichol Magnificently articu-lates a future vision of health, a vision every executive should experience."

SAN BANERJEE,

HEAD OF CONSUMER DIGITAL SOLUTIONS, HUMANA

Made in United States
Troutdale, OR
11/03/2026

56844498R00110